MATHEMATICAL
MODELING

No. 3

Edited by
William F. Lucas, Claremont Graduate School
Maynard Thompson, Indiana University

Clark Jeffries

Mathematical Modeling
in Ecology
A Workbook for Students

Birkhäuser
Boston · Basel · Berlin

Clark Jeffries
Department of Mathematical Sciences
Clemson University
Clemson, SC 29634-1907
USA

Library of Congress Cataloging-in-Publication Data
Jeffries, Clark.
   Mathematical modeling in ecology: a workbook for students.
   (Mathematical modeling ; 3)
   Bibliography: p.
   Includes index.
   1. Ecology—Mathematical models.   I Title.
II. Series: Mathematical modeling ; no. 3.
QH541.15.M3J44   1988   574.5'072'4   88-7955

Printed on acid-free paper

ISBN 0-8176-3421-5
ISBN 3-7643-3421-5

Camera-ready copy prepared by the author.
Printed and bound by R.R. Donnelley and Sons, Harrisonburg, Virginia.
Printed in the USA.

9 8 7 6 5 4 3 2 1

respectfully dedicated to my teacher
Professor H. Gray Merriam
Landscape Ecologist, Carleton University, Ottawa

PREFACE

Mathematical ecology is the application of mathematics to describe and understand ecosystems. There are two main approaches. One is to describe natural communities and induce statistical patterns or relationships which should generally occur. However, this book is devoted entirely to introducing the student to the second approach: to study deterministic mathematical models and, on the basis of mathematical results on the models, to look for the same patterns or relationships in nature.

This book is a compromise between three competing desiderata. It seeks to: maximize the generality of the models; constrain the models to "behave" realistically, that is, to exhibit stability and other features; and minimize the difficulty of presentations of the models. The ultimate goal of the book is to introduce the reader to the general mathematical tools used in building realistic ecosystem models. Just such a model is presented in Chapter Nine. The book should also serve as a stepping-stone both to advanced mathematical works like *Stability of Biological Communities* by Yu. M. Svirezhev and D. O. Logofet (Mir, Moscow, 1983) and to advanced modeling texts like *Freshwater Ecosystems* by M. Straskraba and A. H. Gnauch (Elsevier, Amsterdam, 1985).

In principle, mathematical models in ecology play a role analogous to the role of mathematical models in other disciplines, that is, prediction. In practice, however, ecosystem models are at least as important as frameworks in which to organize scientific thought about ecology. This is in part due to the exceedingly intricate nature of the subject. There is no hope that models based on highly aggregated versions of nature can emulate the predictive power of, say, mechanics or electrodynamics. But by studying mathematical modeling, a student of ecology can nonetheless learn to express the concept of ecosystem in terms of compartments, systems, and flows.

Reading this book should also give the student sufficient background in modeling to understand specific theorems in mathematical ecology. I feel these theorems amount to a foundation for the scientific construction or critical analysis of ecosystem models. Again and again the theorems refer to the qualitative patterns of energy flow in ecosystems as well as certain quantitative aspects of that flow. The point of learning the theorems is being able to focus on the relatively few aspects of ecosystems which determine stability. Thus it is not necessary (and, of course, generally not possible) to specify all details of energy flow functions in an ecosystem model, to account for stability.

Although mathematical modeling tools are not easy to master, the plan of each chapter is to start simply. Only a background in the elements of calculus and linear algebra as normally taught in introductory university courses is assumed. Chapter Two also assumes some knowledge of programming and access to a computer.

I am convinced that understanding the main ideas in this book is a feasible goal for a motivated student who has passed introductory courses in calculus and linear algebra. To such a student who nonetheless does not feel especially comfortable with mathematics, I say, give this book a chance. If you are prepared to exert yourself, you can learn from this book the mathematical side of ecosystem modeling. And while many of the problems amount merely to drill in mechanical skills, it is essential to acquire those skills to form correct and convincing opinions about mathematical modeling. So the beginner should work every problem. For my part, I have tried to write this book suggestively to bring out mathematical ideas you already know about, albeit perhaps not in the context of a mathematical formalism.

I thank Professor P. van den Driessche and Mr. Andrew Clark for ideas and helpful criticism. Mr. Clark's review of an early manuscript was supported by National Science and Engineering Research Council of Canada operating grant A-8965.

C. J.

TABLE OF CONTENTS

Chapter Nine-Sequencing Energy Flow Models to Account for Shortgrass Prairie Energy Dynamics

CHAPTER ONE
AN INTRODUCTION TO DYNAMICAL SYSTEMS AS MODELS

1.1 Ecosystem Development in Terms of Ecology

Let us describe in nonmathematical terms a trajectory of an ecosystem.

Consider the consequences of an early summer forest fire in a Canadian Shield boreal forest of black spruce (*Picea mariana*) with some jack pine (*Pinus banksiana*). With heat activated seed dispersal mechanisms, notably in pine, the next forest is born immediately. The lifespan of a forest, that is, the cycle length from fire to fire, is typically a century. The area of land burned in a fire can vary from less than a hectare to several square kilometers. All forest fires are inherently limited in size; the limits are established by local forest type and age, recent and current weather, and terrain factors such as lakes, ridges of exposed rock, bogs, and sun orientation of faces of wooded hills. The thermal updraft caused by heat released in a large, hot fire can result in local peripheral winds blowing into the fire, and these cooling winds can at least temporarily limit the rate at which the fire grows. On occassion singed pinecones can be dispersed over tens or even hundreds of meters in the strong, turbulent winds of a large fire.

Prompt human intervention at the beginning of a fire can greatly reduce the ultimate size of the burned area. However, regions with a long history of successful fire suppression tend to be susceptable to very large fires. This is, of course, due to the accumulation of dead, dry biomass and to the natural susceptibility of a mature spruce forest to "crown fires," fires propagated at high speed by infrared radiation.

The new forest will grow from the seeds of spruce and pine and the seeds or roots of white birch (*Betula papyrifera*), trembling aspen (*Populus tremuloides*), willow (*Salix* spp.), pin cherry (*Prunus pensylvanica*), and speckled alder (*Alnus rugosa*). Various herbs and shrubs will survive the fire as roots, especially Labrador tea (*Ledum latifolium*; also called *L. groenlandicum*) and blueberry (certain *Vaccinium* species, especially *V. myrtilloides*).

Mosses not burned in the fire may wither from greatly increased light penetration and desiccation, to be replaced by mosses more tolerant of high temperatures, high levels of insolation, and low relative humidity.

Hectares of scorched, blackened pads of moss may prove totally unsuitable for colonization by any visible plants for a decade or more. By contrast, adjacent patches burned down to mineral soil can be colonized promptly and thickly by aspen, willow, cherry, and pine seedlings mixed with fireweed (*Epilobium augustifolium*) and the fumitory *Corydalis sempervirens*. Changes in soil chemistry apparently limit the major bloom of corydalis to the first two or three years following the fire.

Thalloid liverworts can form dense colonies over low, protected patches of clayey soil. Many other local plant distributions will be determined by drainage factors.

There will, of course, be an accompanying sequence of density changes in animal populations. Sawyer beetle larvae (*Monochamus* spp.) will bore into spruce and pine wood within a month of the fire. (Only needles, small branches, and the surface of the bark are burned; the cambium, sapwood, and heartwood are not affected immediately.) In late summer woodpecker activity will greatly increase and remain at a high level until the bark splits on standing dead boles and the boles become desiccated. Yellow-shafted flickers (*Colaptes auratus*), black-backed three-toed woodpeckers (*Picoïdes arcticus*), and northern three-toed woodpeckers (*Picoïdes tridactylus*) can be especially well represented. Snowshoe hare (*Lepus americanus*) will browse willow and birch suckers within a month of the fire. In subsequent years increasing hare populations will lead to increasing populations of predators such as great horned owl (*Bubo virginianus*) and lynx (*Lynx canadensis*).

During wind storms in the decades following the fire, tens or hundreds of burned trees per hectare will be blown down. The effect of roots being ripped up in the storms amounts to a sort of cultivation of the soil. Boreal forest soils can be highly developed podsols, soils which are highly acidic and which are distinguished by a severely leached, ash-colored upper horizon. Likely the cultivation associated with the overturning of trees is a significant long-term factor in maintenance of soil fertility.

Seasonal weather is an overwhelming factor in the boreal forest. Photosynthesis is turned on in daylight in spring and turned off in early fall. There is a terrific input of energy into biomass in summer and a gradual loss in virtually all ecosystem compartments over the remainder of a year. Insects and insectivores are especially closely synchronized with summer.

All this is an example of ecosystem development; for a general discussion of ecosystem development, see [O, Chapter 9].

Now suppose that after the initial fire some additional disturbance takes place. Suppose lightning strikes the standing bole of a burned tree and starts a small, brief fire. This perturbation would change very slightly the overall pattern of changing energy densities. The process of ecosystem development would proceed, and eventually the more recently disturbed area would be indistinguishable. Thus the concept of stability of the developing forest system is suggested.

(It should be emphasized that the above account is merely a sketch of some of the phenomena associated with the boreal forest ecosystem. There is an immense literature on boreal forest succession following a fire and a detailed model would be well outside the limitations of this book. Generally we shall consider highly simplified ecosystem models which suggest the principles of energy flow modeling in spite of their brevity.)

A fundamental problem of ecology is how to account for stable ecosystem development in terms of a model based on energy flow. This is the central problem to be studied in this book. The language of the study is mathematics. The purpose of this chapter is to introduce some of that language.

1.2 State Space, or How to Add Apples and Oranges

Everyone knows intuitively about space, although *space*, *point*, *energy*, and other intuitively appreciated concepts are actually difficult to define. In fact, university mathematics and physics courses leave such fundamental notions undefined. Nonetheless, at the ecological level of organization, the amount of "stuff" (number of individuals, fixed carbon, fixed nitrogen, "energy") in an "ecosystem compartment" is often the criterion by which the compartment is described. For example, in the boreal forest we might have 10 snowshoe hares and .1 great horned owls per square kilometer. We might abbreviate this situation as the pair of numbers (10,.1). This amounts to a point in *two-dimensional space*; we find ourselves already in two-dimensional geometry.

Now without doubt the reader is keen to learn how to add apples and oranges, a task frequently said to be impossible. Suppose we have three apples and four oranges, which we write as (3,4). Suppose we add to this one apple and two oranges, (1,2). The result is four apples and six oranges, (4,6). This is how a modeler adds apples and oranges.

The abstraction of (hares, owls) or (apples, oranges) is the point (x,y) in two-dimensional space. The abstraction of, say, (hares, owls, foxes) is the point (x,y,z) in three-dimensional space. Some ecological models have tens of compartments. The abstraction of the state of a model with n compartments is the point $(x_1,x_2,x_3,...,x_n)$, an *n-dimensional vector* in *n-dimensional space*.

Some people walk about wondering, "What is n-dimensional space, really?" It is nothing more or less than a formal way to think of numbers n at a time, until we assign in a particular model a meaning to each place or component.

So the reader should get used to saying "n-dimensional space." So long as he knows it only means numbers taken n at a time, in a meaningful order, he can be confident he knows what he is talking about.

Now a certain part of n-dimensional space, namely the part where each component of each n-vector is positive, is called the *positive orthant*. Likewise the part where each component is nonnegative is the *nonnegative orthant*. In ecology, models generally only make sense in the positive orthant or possibly the nonnegative orthant. The portion of n-dimensional space wherein a model makes sense is called the *state space* of the model.

To illustrate all this, let us now associate a particular meaning to the components of 5-vectors in the nonnegative orthant of 5-dimensional space. In a model of the nitrogen budget of a northern hardwood forest, let 5-vectors be:

(bound nitrogen in living biomass above ground,
bound nitrogen in living biomass below ground,
bound nitrogen in forest floor,
bound nitrogen in mineral soil,
available soil nitrogen)

At a particular time there would be particular numbers (kg/ha, say) associated with each *vector component* (= model compartment). Thus the state of the model is represented by a *state vector*.

No doubt the reader has encountered field study problems and techniques and so has some feeling for how the above five numbers would be measured at a particular site. If the above 5-vector were measured at ten sites, the data would amount to ten 5-vectors. The averages of the values over the ten sites would be yet another 5-vector.

Now we come to the central tool of this book, the tool with which we shall model ecosystems in terms of energy flow: **the dynamic model (a model which changes with time)**. Such a model must specify not only ecosystem compartments (and so meaning for the components of n-vectors) but also mathematical rules or functions which determine the dynamic development of the system.

Consider, for example, the following model of bacteria growing in a laboratory medium: at noon on a certain day there are 100 bacteria present in the medium; every hour thereafter the number of bacteria doubles. Here we have only one model compartment, a 1-vector, the current number of bacteria. In mathematics such a model is represented by the equation

$$x(t+1) = 2x(t) \qquad \text{(with } x(0) = 100)$$

where x is the variable denoting number of bacteria and t is time measured in hours. As the reader well knows, the numbers of bacteria hour by hour would then be: 100, 200, 400, 800, and so on. Such a sequence of model states is called a *trajectory* for the model.

A *perturbation* of a dynamic model is a brief or instantaneous change caused by external factors to which the model subsequently responds. We might add 100 additional bacteria to the above model at t = 0. Thus we would have

$$x(t+1) = 2x(t) \qquad \text{(with } x(0) = 200)$$

The resulting trajectory would be: 200, 400, 800, 1600, and so on.

So far everything is clear, right? For a realistic ecological model with many compartments, a perturbation would be just like adding apples and oranges, in other words, component added to component in the initial state vector. The model would then exhibit certain changes, and if the model were carefully and scientifically designed, those changes would be reflected in nature. Moving back to the northern hardwood forest model, we might add to the observed state vector the perturbation (0,0,0,0, nitrogen added as available soil nitrogen). The addition of such fertilizer would still be just adding apples and oranges, in other words, component added to component. Presumably such an addition would have mathematical consequences for the model and real consequences for the forest. The general purpose of creating energy flow models is to try to use the former to predict the later.

Let's back up to the idea of function. A *function* is a mathematical machine. If we give the function a number, it gives us a number. Slightly fancier functions can consume several numbers at once and produce several numbers. For example, a certain function might consume the 3-vector (latitude, longitude, day of year) and produce the 2-vector (time of sunrise, length of daylight period). Another function might consume the 5-vector (small rodent population density, snowshoe hare population density, ruffed grouse population density, spruce grouse population density, time of year) and produce the 4-vector consisting of the average number of kills per mink per day involving each of the four groups of prey.

Much of the hard work in ecological model building amounts to determining such functions.

1.3   Dynamical Systems as Treasure Hunts

Now what exactly does an ecosystem model look like? It looks like the clues in a treasure hunt. Of course, everyone knows about treasure hunts in which one finds and analyzes (computes the value of) a clue to obtain directions to another place, whereupon another clue is found, and so on.

Suppose we regard a piece of graph paper as a map, north to the top. Fix an origin--the 2-vector (0,0)--in the middle of the graph paper and consider the following list of locations (2-vectors or state vectors on the graph paper) and associated clues.

location	clue
(1,1)	go 1 east, 0 north
(1,2)	go 0 east, -1 north (1 south)
(1,3)	go 0 east, -1 north
(2,1)	go 0 east, 1 north
(2,2)	go 0 east, 0 north
(2,3)	go -1 east, 0 north
(3,1)	go 0 east, 1 north
(3,2)	go 0 east, 1 north
(3,3)	go -1 east, 0 north

Let us try starting at the point (3,1). After the first move we find ourselves at (3,2), of course. After eight moves we reach the state (2,2). In other words, at one move a second we arrive at (2,2) after eight seconds. Where would we be after 100 additional seconds?

This simple exercise is an example of a *difference equation dynamical system*. At each *time step* Δt (here one second) the system jumps a "difference" through space. The difference is given by the clue list. Each clue or difference is given as a function of location, location expressed as a 2-vector.

Let's translate all this into mathematics in order to develop a general notation for difference equation dynamical systems.

If the location or state is (x,y), let the difference (what is added to (x,y) to get to the next state) be

$$(f(x,y)\Delta t, \; g(x,y)\Delta t)$$

at each time step Δt. Thus $f(x,y)\Delta t$ is the number added to the first component x of (x,y) to get to the next state. Likewise g(x,y) Δt is the number added to the second component y of (x,y) to get to the next state.

Now here is a question to test the reader's understanding of the difference (f(x,y)Δt, g(x,y)Δt): what is f(1,1)? To answer this, go back to the clue list. At location (1,1) we are to "go 1 east, 0 north." Thus the amount added to the first component is 1. Taking Δt = 1 (one second), f(1,1) must be 1. Likewise g(1,1) = 0. So (f(1,1)·1, g(1,1)·1) = (1,0). The reader should be able to find (f(x,y), g(x,y)) for any location (x,y) in the list.

The general mathematical formulation for a *two-dimensional difference equation dynamical system* is

$$x(t+\Delta t) = x(t) + f(x,y)\,\Delta t$$
$$y(t+\Delta t) = y(t) + g(x,y)\,\Delta t$$

(Understanding these two equations is equivalent to understanding a simple form of the fundamental idea of the mathematical idea of modeling, so exert yourself!) The first equation means: to find the next x value, that is, the value of x at time t+Δt, add to the old x value, namely x(t), the product of the local rate of change (given by a clue list or the mathematical equivalent) and the time step (one second, one day, or whatever other time interval is suitable at the ecosystem level of organization). It is crucial to understand the parts of these equations: f(x,y)Δt is the difference or change; Δt is the amount of time in each step; and f(x,y) is the change in x per time step at the state (x,y).

Similarly, the new value of y at time t+Δt is obtained by adding to the current value y(t) the product of the local rate of change per time step g(x,y) and the time step Δt itself.

The above list of clues could be written in function form as follows:

$$f(x,y)\Delta t = \begin{cases} +1 \text{ at } (1,1) \\ -1 \text{ at } (2,3) \text{ and } (3,3) \\ 0 \text{ at other points} \end{cases}$$

$$g(x,y)\Delta t = \begin{cases} +1 \text{ at } (2,1), (3,1), \text{ and } (3,2) \\ -1 \text{ at } (1,2) \text{ and } (1,3) \\ 0 \text{ at other points} \end{cases}$$

The sequence of points we found earlier by following the clues is a trajectory for the system with Δt = 1 second starting at t = 0 at the initial state (3,1). Starting at any other initial point in the list generally leads to another trajectory. In the case of the above clue list all trajectories eventually become the rather repetitious sequence (2,2), (2,2), (2,2), ... The trajectory starting at (2,2) is called a *constant trajectory* because it never changes with time, that is, the rates of change f(2,2) and g(2,2) are both 0. Moreover, this particular constant trajectory is called an *attractor trajectory* because after a certain amount of time determined by initial conditions any trajectory will be "arbitrarily close" to (2,2) ("within epsilon," as mathematicians enjoy saying, using the standard measure of distance in 2-dimensional space). In fact, any other trajectory will wind up right up on top of (2,2)! In a sense, the constant attractor trajectory (2,2) is the treasure of the treasure hunt clue list, the system equations.

Suppose we have instead of a clue list the mathematical rules: $\Delta t = .001$; $f(x,y) = 200x - x^2 - 10xy$; and $g(x,y) = xy - 10y^2$; or the equivalent as equations:

$$x(t+.001) = x(t)+[200x(t)-x^2(t)-10x(t)y(t)](.001)$$
$$y(t+.001) = y(t)+[x(t)y(t)-10y^2(t)](.001)$$

The reader should be able to show that (100,10) is a constant trajectory for this system. Next consider the trajectory for this system starting at $t = 0$, $x(0) = 80$, $y(0) = 15$. It must be possible for a determined, careful person to calculate by hand the values of x and y along the trajectory for, say, 50 time steps, up to time = .050. No doubt the reader would agree that it would be much better to give such a task to a microcomputer, or at least a programmable calculator. (A program for computing the trajectory on a Hewlett Packard 41CV calculator is given at the end of this chapter.) At any rate, some typical values of the trajectory are approximately:

$(x(0), y(0)) = (80,15)$
$(x(.050), y(.050)) \cong (100.1128, 10.0375)$
$(x(.100), y(.100)) \cong (100.0024, 10.0001)$
$(x(.150), y(.150)) \cong (100.0000, 10.0000)$

A sketch of points of the trajectory is shown in Fig. 1.1. The points are connected for the sake of visual clarity.

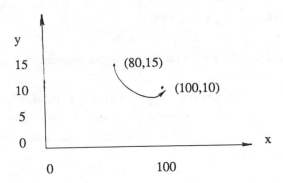

Figure 1.1. A trajectory for the two-dimensional difference equation dynamical system:

$x(t+\Delta t) = x(t)+[200x(t)-x^2(t)-10x(t)y(t)]\Delta t$

$y(t+\Delta t) = y(t)+[x(t)y(t)-10y^2(t)]\Delta t$

This model is a simplistic version of prey (x) and predator (y) dynamics. Figure 1.1 obviously suggests a sort of *stability*: the constant trajectory (100,10) is the treasure of the treasure hunt starting at (80,15). There is a perverse element to the mathematics, however, in that the treasure is never actually reached (unless we are numerically extremely lucky). Sooner or later the modeler must say that the trajectory for all practical purposes has reached the treasure. It is at this philosophical stage that the mathematician (who, nonetheless, goes around believing that .999999... equals 1) feels uneasy. The modeler, moreover, might compute other trajectories starting near (100,10), say, starting at (115,5) or (120,20). Each such trajectory will likewise seem to approach asymptotically the constant trajectory (100,10), so the modeler is tempted to assert that the equations are stable in some sense. The mathematician feels worse.

Fortunately there is a line of reasoning which will satisfy the mathematician that all trajectories starting in the positive orthant <u>must</u> asymptotically approach the constant trajectory (100,10). The theory will be explained in subsequent chapters.

The reader can no doubt appreciate that ecosystem models which are at all realistic must have tens or even hundreds of compartments. In a model of large dimension there would not be enough letters in the English alphabet or any other Western alphabet to assign one letter to each compartment as its algebraic label. So we resort to subscripts, numbers attached to a general component symbol x. We write $x = (x_1, x_2, x_3, ..., x_n)$ and so x becomes a shorthand for the whole n-vector. This leads to the general difference equation for x_1:

$$x_1(t+\Delta t) = x_1(t) + f_1(x_1(t), x_2(t),...,x_n(t))\Delta t$$

The reader should be able to write out the analogous equations for $x_2(t+\Delta t)$, $x_3(t+\Delta t),...,x_n(t+\Delta t)$.

Much of mathematics amounts to the creation and manipulation of mathematical shorthand. Therefore we might as well also abbreviate the *rate of change functions* $(f_1, f_2, f_3,...,f_n) = f$ so the whole *difference equation dynamical system* can be written as

$$x(t+\Delta t) = x(t) + f(x(t))\Delta t:$$

Let's consider another example. Let $n = 4$ and

$$f(x) = (1000x_1 - 10x_1x_2, 10x_1x_2 - 500x_2x_3 - 500x_2x_4, 50x_2x_3 - 500x_3^2, 30x_2x_4 - 300x_4^2)$$

This $f(x)$ is associated with the difference equation dynamical system

$$x_1(t+\Delta t) = x_1(t) + [1000x_1 - 10x_1x_2]\Delta t$$
$$x_2(t+\Delta t) = x_2(t) + [10x_1x_2 - 500x_2x_3 - 500x_2x_4]\Delta t$$
$$x_3(t+\Delta t) = x_3(t) + [50x_2x_3 - 500x_3^2]\Delta t$$
$$x_4(t+\Delta t) = x_4(t) + [30x_2x_4 - 300x_4^2]\Delta t$$

(Each x_i inside square brackets $[\cdot]$ actually means $x_i(t)$, of course.) The reader should be able to show that $(1000,100,10,10)$ is a constant trajectory for this system. Again there is a limited interpretation of these equations as a model of population dynamics: compartment 1 is an autotroph species eaten by the species in compartment 2; in turn the species in compartment 2 is preyed upon by the species in compartments 3 and 4. Can the reader see this reflected in terms in $f(x)$?

Outside factors are important in many ecosystem models. Temperature, relative humidity, sunlight intensity and duration, precipitation, and pollution inputs are examples of such factors. So it would be desirable in some circumstances to allow ourselves the freedom to incorporate time dependent factors explicitly and directly in the rate of change function f. In such cases we would use the *time dependent difference equations*

$$x(t+\Delta t) = x(t) + f(x(t),t)\Delta t.$$

A hypothetical example of such a system is given in problem 20.

REFERENCE

[O] Eugene P. Odum, *Fundamentals of Ecology*, Saunders, Philadelphia, 1971.

APPENDIX

The following program for a Hewlett Packard 41CV calculator can be used to find trajectories for the difference equation dynamical system

$$x(t+\Delta t) = x(t)+[200x(t)-x^2(t) -10x(t)y(t)]\Delta t$$
$$y(t+\Delta t) = y(t)+[x(t)y(t) -10y^2(t)] \Delta t$$

01	LBL ᵀDYN SYS	18	-10	35	RCL 02	52	INT
02	ᵀINITIAL X?	19	*	36	$x / 2$	53	RCL 05
03	PROMPT	20	RCL 01	37	-10	54	-
04	STO 01	21	$x / 2$	38	*	55	$x = 0$?
05	ᵀINITIAL Y?	22	-	39	+	56	XEQ ᵀRDOT
06	PROMPT	23	RCL 01	40	RCL 04	57	ISG 00
07	STO 02	24	200	41	*	58	GTO ᵀDE
08	ᵀTIME STEP?	25	*	42	RCL 02	59	LBL ᵀRDOT
09	PROMPT	26	+	43	+	60	RCL 01
10	STO 04	27	RCL 04	44	STO 02	61	PSE
11	ᵀISG NUMBER?	28	*	45	RCL 03	62	RCL 02
12	PROMPT	29	RCL 01	46	STO 01	63	PSE
13	STO 00	30	+	47	RCL 00	64	RTN
14	LBL ᵀDE	31	STO 03	48	INT	65	END
15	RCL 01	32	RCL 01	49	.1		
16	RCL 02	33	RCL 02	50	*		
17	*	34	*	51	STO 05		

This program asks for initial x, initial y, and time step values. The subroutine DE (difference equation) calculates new x and y values at each time step. The subroutine RDOT (read out) displays every tenth value of the trajectory. A time step no larger than .001 will ensure numerical stability. Using an ISG (increment stepping number) of .04901 yields 50 time steps. Using ISG = .99901 yields 1000 time steps.

PROBLEMS

1. Let

$x = (1, 2, 3)$
$y = (-1, 0, -4)$
$z = (0, 2, 0)$

Find the sums x+y, x+z, and x+y+z.

2. Let

$x = (0, 0, 1, 2, 0)$
$y = (0, 1, 0, 0, 2)$
$z = (1, 0, 0, 0, 3)$

Find the sums x+y, x+z, and x+y+z.

3. Scalar multiplication is a way to multiply a number (scalar) and a vector. An example of scalar multiplication is multiplication of f and Δt in a difference equation dynamical system. Generally the scalar product of s (a number) and $x = (x_1, x_2, ..., x_n)$ is $sx = (sx_1, sx_2, ..., sx_n)$. Using the 3-vectors in problem 1, find 2x, 3x+2y, and -1x+y-2z.

4. Using the 5-vectors in problem 2, find 3x, -3x+2y, -1x-1y+4z.

5. The Pythagorean Theorem in two-dimensional space states that the distance $d(x,y)$ from $x = (x_1, x_2)$ to $y = (y_1, y_2)$ is the square root of $(x_1-x_2)^2+ (y_1-y_2)^2$, that is,

$$d(x,y) = \left[\sum_{i=1}^{2} (x_i - y_i)^2 \right]^{1/2}$$

Let $x = (1,0)$, $y = (5,4)$, and $z = (3,6)$. Is y closer to x or z?

6. The Pythagorean Theorem may be extended to define the distance $d(x,y)$ between any two points (n-vectors) x and y in n-dimensional space. The distance is

$$d(x,y) = \left[\sum_{i=1}^{n} (x_i - y_i)^2\right]^{1/2}$$

Find the distance $d(x,y)$ from $x = (1,0,2,2,0)$ to $y = (2,0,0,2,2)$. Find the distance $d(y,x)$ from y to x.

7. Using the 3-vectors in problem 1, find $d(x,y)$, $d(y,z)$, and $d(x,z)$.

8. Using the 5-vectors in problem 2, find $d(x,y)$, $d(y,z)$, and $d(x,z)$.

9. The length of an n-vector x is defined to be the distance $d(x,0)$ from x to the origin $0 = (0,0,0,...,0)$. Of the three 5-vectors in problem 2, which is the longest?

10. In 6-dimensional space let

$x = (1, 3, -2, 0, -2, 1)$
$y = (0, 0, 1, 2, 1, 2)$
$z = (-2, 0, -3, -6, 1, 1)$

Calculate $d(x,y)+d(y,z)$ and $d(x,z)$. Find $d(x+2y,z)$.

11. Using the 6-vectors in problem 10, show that the product of x_2 and y_4 is 6. Find the product $x_2 y_5 z_6$. Find $(x_5 y_3 + y_2 z_4 + x_3 y_3 y_3 z_3)z_1$.

12. In 3-dimensional space let a function $f(x)$ be defined by $f(x) = x_1 + x_2 + 2x_3$. (Hence $f(y) = y_1 + y_2 + 2y_3$, $f(z) = z_1 + z_2 + 2z_3$, and so on.). Using

$x = (1, 2, 3)$
$y = (-1, 0, -4)$
$z = (0, 2, 0)$

calculate $f(x)$, $f(x+y)$, and $f(x+y+z)$.

13. Let $g(x)$ be a function in 3-dimensional space defined by $g(x) = x_1 - 2x_1 x_2 + x_3^2$.

Using x, y, z of problem 12, calculate g(x), g(y), and g(y+z).

14. A difference equation dynamical system in 2-dimensional space is defined by the equations

$$x_1(t+\Delta t) = x_1(t)+[-x_1(t)-x_2(t)]\Delta t$$
$$x_2(t+\Delta t) = x_2(t)+x_1(t)\Delta t$$

Let $\Delta t = 1$ (that is, the time step is one unit), and compute values of the trajectory starting at $t = 0$ at (1,0). Graph the points of the trajectory up to $t = 7$. Evaluate (x(60),y(60)) using the graph.

15. Consider the 1-dimensional system

$$x(t+\Delta t) = x(t)+[-.5x(t)]\Delta t$$

If $x(0) = 10$ and $\Delta t = 1$, what is x(4)? What value will x(t) approach as t becomes arbitrarily large?

16. Using algebra one can show that the general trajectory value for

$$x(t+\Delta t) = x(t)+[-.5x(t)]\Delta t$$

is $x(p\Delta t) = x(0)(1 - .5\Delta t)^p$ where $p = 0,1,2,3,...$. Assuming $x(0) = 10$, use a calculator to find x(4) with $\Delta t = .0001, .001, .01, .1, .5, 2$, and 4. What do the results suggest about using large Δt or small Δt?

17. Consider the 1-dimensional time driven system

$$x(t+\Delta t) = x(t)+[-.5x(t)+\sin(.5\pi t)]\Delta t$$

Let $\Delta t = 1$ and $x(0) = 10$. Determine x(4).

18. Consider the 3-dimensional system

$x_1(t+\Delta t) = x_1(t)+x_1(t)[-x_1(t)-x_2(t)+1100]\Delta t$
$x_2(t+\Delta t) = x_2(t)+x_2(t)[.1x_1(t)-.9x_2(t)-x_3(t)]\Delta t$
$x_3(t+\Delta t) = x_3(t)+x_3(t)[.1x_2(t)-x_3(t)]\Delta t$

Find a constant trajectory for this system in the positive orthant.

19. For the system in problem 18, let $x(0) = (1000,100,10)$ and $\Delta t = .001$. Find $x(1)$. Given sufficiently small Δt, all trajectories for the system which start near the constant trajectory asymptotically approach it with time. As will be explained in Chapter 4, this model may be associated with a simplistic "predation community" in which x_1 is an autotroph compartment, x_2 is an herbivore compartment, and x_3 is a carnivore compartment.

20. Consider the 3-dimensional system

$x_1(t+\Delta t) = x_1(t)+x_1(t)[-x_1(t)-x_2(t)+1100+\cos(\pi t)]\Delta t$
$x_2(t+\Delta t) = x_2(t)+x_2(t)[.1x_1(t)-.9x_2(t) -x_3(t)]\Delta t$
$x_3(t+\Delta t) = x_3(t)+x_3(t)[.1x_2(t)-x_3(t)]\Delta t$

This system differs from that in problem 18 only by the "$\cos(\pi t)$" term in the equation for $x_1(t+\Delta t)$. Let $\Delta t = .001$ and $x(0) = (1000,100,10)$. Calculate $x(.001)$. In this case the autotroph component x_1 is time driven in that its "intrinsic growth coefficient" is $1100+\cos(\pi t)$ and so is cyclic. This small time dependence causes typical trajectories of the whole system to approach and enter a small regionin 3-dimensional space around $(1000,100,10)$ and thereafter oscillate in an irregular manner forever.

21.   Consider the 5-dimensional system

$x_1(t+\Delta t) = x_1(t)+[-x_1(t)+x_2(t)]\Delta t$
$x_2(t+\Delta t) = x_2(t)+[-x_1(t)-x_2(t)+x_3(t)]\Delta t$
$x_3(t+\Delta t) = x_3(t)+[x_2(t)-x_3(t)-x_4(t)]\Delta t$
$x_4(t+\Delta t) = x_4(t)+[x_3(t)-x_4(t)-x_5(t)]\Delta t$
$x_5(t+\Delta t) = x_5(t)+[x_4(t)-x_5(t)]\Delta t$

Describe the trajectory which starts at $(1,0,0,0,1)$ with $\Delta t = .5$ for times $.5, 1, 1.5, 2, 2.5, 3, 3.5, 4$.

22. Consider the 8-dimensional system

$x_1(t+\Delta t) = x_1(t)+[x_1(t)-x_1(t)x_2(t)]\Delta t$
$x_2(t+\Delta t) = x_2(t)+[2x_1(t)x_2(t)- x_2(t)- x_2(t)^2]\Delta t$
$x_3(t+\Delta t) = x_3(t)+[x_3(t)-x_3(t)x_4(t)]\Delta t$
$x_4(t+\Delta t) = x_4(t)+[2x_3(t)x_4(t)- x_4(t)- x_4(t)^2]\Delta t$
$x_5(t+\Delta t) = x_5(t)+[x_2(t)+x_3(t)-2x_5(t)x_6(t)]\Delta t$
$x_6(t+\Delta t) = x_6(t)+[2x_5(t)x_6(t)- x_6(t)- x_6(t)^2]\Delta t$
$x_7(t+\Delta t) = x_7(t)+[x_6(t)-x_7(t)x_8(t)]\Delta t$
$x_8(t+\Delta t) = x_8(t)+[x_7(t)x_8(t)-x_8(t)]\Delta t$

Verify that $(1,1,1,1,1,1,1,1)$ is a constant trajectory for this system. Then let $\Delta t =$ .25 and $x(0) = (2,1,1,1,1,1,1,1)$ and verify that $x(1)$ is approximately $(.871, 1.88, 1, 1, 1.44, 1.22, 1.02, 1)$.

23. In problem 22, the component values x_1 and x_2 comprise a subsystem which drives the rest of the system but is not driven by the rest of the system. In other words, the future values of the other system components depend upon $x_1(t)$ and $x_2(t)$ but $x_1(t)$ and $x_2(t)$ depend only on $x_1(0)$ and $x_2(0)$. If $x_1(0) = x_2(0) = 1$ and Δt t= .001, find x_1 (1000) and x_2 (1000), regardless of $x_3(0)$, $x_4(0)$, $x_5(0)$, $x_6(0)$, $x_7(0)$, or $x_8(0)$.

24. Let $f = (f_1,f_2,...,f_8)$ be the system functions in problem 22. Find f_4 (1,1,2,1,4,1,1,2).

ANSWERS TO PROBLEMS

1. $x+y = (0,2,-1)$; $x+z = (1,4,3)$; $x+y+z = (0,4,-1)$

2. $x+y = (0,1,1,2,2)$; $x+z = (1,0,1,2,3)$; $x+y+z = (1,1,1,2,5)$

3. $2x = (2,4,6)$; $3x+2y = (1,6,1)$; $-1x+y-2z = (-2,-6,-7)$

4. $3x = (0,0,3,6,0)$; $-3x+2y = (0,2,-3,-6,4)$; $-x-y+4z = (4,-1,-1,-2,10)$

5. $d(x,y) = 5.657 > d(y,z) = 2.828$

6. $d(x,y) = d(y,x) = 3$

7. $d(x,y) = 7.550$; $d(y,z) = 4.583$; $d(x,z) = 3.162$

8. $d(x,y) = 3.162$; $d(y,z) = 1.732$; $d(x,z) = 3.873$

9. $d(x,0) = 2.236$; $d(y,0) = 2.236$; $d(z,0) = 3.162$

10. $d(x,y)+d(y,z) = 14.964$; $d(x,z) = 8$; $d(x+2y,z) = 12$

11. $x_2y_4 = 3 \cdot 2 = 6$; $x_2y_5z_6 = 3$; $(x_5y_3+y_2z_4+x_3y_3z_3)z_1 = -8$

12. $f(x) = 9$; $f(x+y) = 0$; $f(x+y+z) = 2$

13. $g(1,2,3) = 6$; $g(-1,0,-4) = 15$; $g(-1,2,-4) = 15$

14. The points of the trajectory occur cyclically as $(1,0)$, $(0,1)$, $(-1,1)$, $(-1,0)$, $(0,-1)$, $(1,-1)$, $(1,0)$,.... . Hence $x(60) = 1$ and $y(60) = 0$.

15. $x(4) = .625$; $x(t)$ approaches 0 as t becomes arbitrarily large.

16.

$t =$	.0001	.001	.01	.1	.5	1	2	4
$x(4) \cong$	1.353	1.353	1.347	1.285	1.001	.625	undefined	-10

The results suggest that using too large a Δt value leads to diverging trajectories and that using too small a t value leads to unnecessarily long calculations.

17. $x(4) \cong .125$

18. In the positive orthant a constant trajectory occurs if and only if

$-x_1-x_2+1100 = 0$
$.1x_1-.9x_2-x_3 = 0$
$.1x_2-x_3 = 0$

The third equation implies $x_2 = 10x_3$ and the second equation then implies $x_1 = 100x_3$. From the first equation $x_3 = 10$, so $x_2 = 100$ and $x_3 = 1000$.

19. $x(1) = x(\text{any } t) = (1000, 100, 10)$

20. $x(.001) = (1001, 100, 10)$

21. $x(0.5) = (.5,-.5,0,-.5,.5)$
      $x(1.0) = (0,-.5,0,-.5,0)$
      $x(1.5) = (-.25,-.25,0,-.25,-.25)$
      $x(2.0) = (-.25,0,0,0,-.25)$
      $x(2.5) = (-.125,.125,0,.125,-.125)$
      $x(3.0) = (0,.125,0,.125,0)$
      $x(3.5) = (.0625,.0625,0,.0625,.0625)$
      $x(4.0) = (.0625,0,0,0,.0625)$
Thus $x(4) = (.0625) x(0)$. As t increases arbitrarily, $x(t)$ asymptotically approaches $(0,0,...,0)$.

22. A programmable calculator can be used to perform the verifications.

23. $x_1(1000) = x_1 \text{ (any } t) = 1; x_2(1000) = x_2 \text{ (any } t) = 1$

24. $f_4 = 2x_3x_4-x_4-x_4^2 = 2$

CHAPTER TWO
SIMPLE DIFFERENCE EQUATION MODELS

2.1 Predator-Prey Difference Equation Dynamical Systems

The rate of energy transfer between ecosystem compartments either does or does not depend upon the current state of the recipient compartment. In the former case the compartments are said to be involved in a *predation interaction*. Dairy cattle grazing grass, mosquitoes drawing blood from humans, ravens comsuming carrion, and soil bacteria digesting detritus are all examples of predator-prey pairs in predation interactions. Energy transfer between ecosystem compartments which is not a predation interaction is called *detritus donation*. The donation of energy from the moss compartment to the peat compartment in a model of a northern bog would be an example of detritus donation. This chapter concerns modeling a predation interaction at the simplest level.

How much prey energy will be transferred to a predator compartment per day (or other suitable time unit)? Obviously there is an element of chance or probability in the answer to such a question. In this section, however, let us deal with average energy transfer rates. (It is actually a tacit but debatable assumption in the general form of the difference equation dynamical system that knowing all current compartment levels--including all population densities--and time of day or year is sufficient to determine any average rate of energy transfer through predation.)

As a very first step, however, let us propose that the rate of predation energy transfer is simply proportional to the product of predator and prey compartment levels. Using a heuristic understanding of classical statistical mechanics, we can reason that in an agitated box containing black marbles and white marbles the rate of collisions between black marbles and white marbles is on the average proportional to the product of the number of white marbles and the number of black marbles. This is called *"collisions" type predation energy flow*.

From the classical studies of A. J. Lotka and Vito Volterra early in the twentieth century, difference equations for the simplified dynamics of prey and predator are

$$x(t+\Delta t) = x(t)+[-Ax^2(t)-Bx(t)y(t)+Ex(t)]\Delta t$$
$$y(t+\Delta t) = y(t)+[Cx(t)y(t)-Dy^2(t)-Fy(t)]\Delta t$$

where $x(t)$ is the prey population, $y(t)$ is the predator population, and A,B,C,D,E,F are all positive constants. (Generally in mathematical modeling the convention is made that "x" and "y" can be taken as mathematical shorthand for "x(t)" and " y(t)." In this chapter only, let us write out x(t) and y(t) in full each time in order to reduce slightly the risk of confusion faced by the nonmathematician reader.) The -Bx(t)y(t) term is thus the loss rate from x(t) due to predation and the +Cx(t)y(t) is the gain rate to y(t) from predation. The $-Ax^2(t)$ and $-Dy^2(t)$ terms represent auto-regulation terms, losses resulting from intraspecific "collisions" involving territorial conflicts, competition among autotrophs for sunlight, intraspecific predation, and so on. The +Ex(t) term is the rate of growth x(t) would enjoy in the absence of predators and auto-regulation, that is, +Ex(t) is the Malthusian or intrinsic growth rate. The -Fy(t) term is the decline rate y(t) would suffer in the absence of prey and auto-regulation.

This model is presented only as a stepping stone to more realistic models. For illustrative purposes we now consider the particular equations

$$x(t+\Delta t) = x(t)+[-.1x^2(t)-2x(t)y(t)+2.1x(t)]\Delta t$$
$$y(t+\Delta t) = y(t)+[2x(t)y(t)-.1y^2(t)-1.9y(t)]\Delta t$$

Note that (x(t), y(t)) = (1,1) is a constant trajectory for this system. Suppose as initial state we have (x(0),y(0)) = (2,1), and suppose $\Delta t = .01$. The first few values for the resulting trajectory are approximately

t	x(t)	y(t)
0	2	1
.01	1.998	1.020
.02	1.995	1.040
.03	1.992	1.061

Continuing through more time values leads to a spiral pattern in two-dimensional space. Roughly speaking, energy sloshes back and forth between the two compartments. The graph in Figure 2.1 shows the pattern of such data points connected by lines for clarity. The total time interval for the curve is 2.23 units.

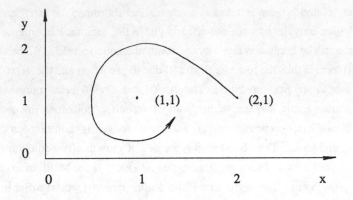

Figure 2.1. A trajectory for a predator-prey model.

As it happens, little qualitative difference exists between $\Delta t = .01$ and smaller Δt values, so in order to conserve calculation time we let $\Delta t = .01$ in the remainder of this section.

The following is a BASIC language program which can be used to generate the trajectory in Fig. 2.1 and any other trajectory for the model.

```
100   DEF FNF1(X,Y)=-0.1*X*X-2*X*Y+2.1*X
110   DEF FNF2(X,Y)=2*X*Y-0.1*Y*Y-1.9*Y
120   INPUT "Enter initial values of x, y",X,Y
130   INPUT "Enter initial value of t",T0
140   INPUT "Enter final value of t",T1
150   INPUT "Enter delta t",S
160   FOR T=T0 TO T1 STEP S
170   PRINT T,X,Y
180   DX=S*FNF1(X,Y)
190   DY=S*FNF2(X,Y)
200   X=X+DX
210   Y=Y+DY
220   NEXT T
230   END
```

Lines 100 and 110 define the rate of change functions. Lines 120 through 150 ask the modeler for initial values and step size. Lines 160 through 220 perform the calculations and print the time and trajectory values. A slightly longer program shown listed in the Appendix of this chapter plots the trajectories of the predator-prey equations.

It is also instructive to imagine trajectories for the predator-prey system in three dimensions: x, y, and time t. A typical trajectory is a curve winding around and becoming ever closer to the straight line consisting of points (1,1,t). Such a curve is shown in Fig. 2.2. As can be shown using the mathematical tools in Chapter Four, all trajectories for all predator-prey models with A,B,...,F>0 and sufficiently small Δt look qualitatively like those in Fig. 2.1 and 2.2, with greater or lesser degrees of "spiraling" as they approach (1,1) or (1,1,t).

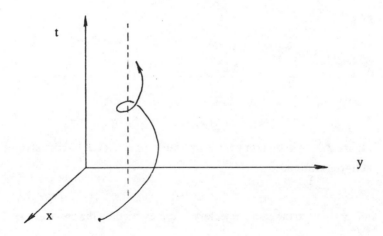

Figure 2.2. A trajectory for a predator-prey model in (x,y,t)-space.

In the model the predation coefficients -2 and +2 are large in magnitude (absolute value) compared to the auto-regulation coefficients -.1 and -.1. It is this fact which accounts for the pronounced spiraling of the trajectories, the phenomenon of both populations repeatedly overshooting the constant trajectory values 1 and 1. Overshooting would not occur in a system with auto-regulation terms of large magnitude. For example, computer simulation of the model

$$x(t+\Delta t) = x(t)+[-2x^2(t)-.1x(t)y(t)+2.1x(t)]\Delta t$$
$$y(t+\Delta t) = y(t)+[2x(t)y(t)-1.9y^2(t)-.1y(t)]\Delta t$$

shows that a typical trajectory looks like that in Fig. 2.3. (The reader should be able to associate each coefficient in the rate of change functions of this model with an ecological interaction.)

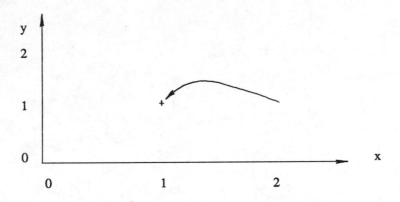

Figure 2.3. A trajectory for a predator-prey model having strong auto-regulation.

In (x,y,t)-space the same trajectory is represented by the curve in Fig. 2.4.

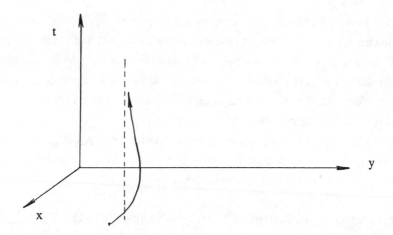

Figure 2.4. A trajectory for a predator-prey model in (x,y,t)-space.

In all of the above it is important to realize that the trajectories are not smooth curves but points in space which are connected by a curve for the sake of visual clarity.

2.2 Probabilistic Limit Cycles

Given all the coefficients, starting values, and Δt, the above difference equation version of the predator-prey relationship yields a unique future trajectory, at least unique up to small round-off errors in actual computer runs. Thus the difference equation version of the predator-prey is deterministic. Nature does not, of course, operate deterministically at the ecosystem level of organization. Rather, she rolls dice at every instant to introduce probability in small or large measure into the running of all real ecosystems. Fluctuations in precipitation, temperature, insolation intensity and duration, relative humidity, and wind speed would all be probabilistic factors in the energy flow rates in ecosystems. Likewise, the predation interactions between species can be characterized by chance, a characterization to which any fisherman or hunter can attest. This section is about mathematically incorporating lumped probabilistic factors into the predator-prey equations.

The incorporation is accomplished as follows. Consider the equations which in the previous section gave rise to spiraling trajectories

$x(t+\Delta t) = x(t)+[-.1x^2(t)-2x(t)y(t)+2.1x(t)]\Delta t$
$y(t+\Delta t) = y(t)+[2x(t)y(t)-.1y^2(t)-1.9y(t)]\Delta t$

In going from each t to t+Δt let us simulate all probabilistic factors concerning x by ordering the computer to: select a random number between -1 and +1; multiply that number by a scaling constant; multiply that product by x(t); and add the result to x(t+Δt). Likewise, the computer will be required at each time step to do the same (with a different random number, generally) in calculating y(t+Δt). A scale constant of .07 will be used because, as experiments have shown, .07 is large enough to have a significant effect on system dynamics and is not large enough to immediately drive the system out of the positive orthant. Thus our probabilistic difference equation dynamical system becomes

$$x(t+\Delta t) = x(t)+[-.1x^2(t)-2x(t)y(t)+2.1x(t)]\Delta t+.07x(t)[2RND(1)-1]$$
$$y(t+\Delta t) = y(t)+[2x(t)y(t)-.1y^2(t)-1.9y(t)]\Delta t+.07y(t)[2RND(2)-1]$$

The randomizing terms are, of course, 2RND(1)-1 and 2RND(2)-1. Here RND(1) and RND(2) are random numbers between 0 and 1 generated by the computer, so 2RND(1)-1 and 2RND(2)-1 are random numbers between -1 and +1.

In the BASIC program we must change lines 200 and 210 to

```
200   X=X+DX+0.07*X*(2*RND(1)-1)
210   Y=Y+DY+0.07*Y*(2*RND(2)-1)
```

The perturbations might be thought of as little kicks which send the trajectory off its spiral track, possibly closer to (1,1), possibly further from (1,1). The trajectory then starts again down another track until another kick occurs. In computer runs the overall effect is that the trajectory generally does not approach (1,1) but rather oscillates around (1,1) in a fairly regular way. A typical run is shown in Fig. 2.5. To prove that the system is not attracted to (1,1), the trajectory was actually started at (1,1). Hence 1, 1 was used in response to line 120.

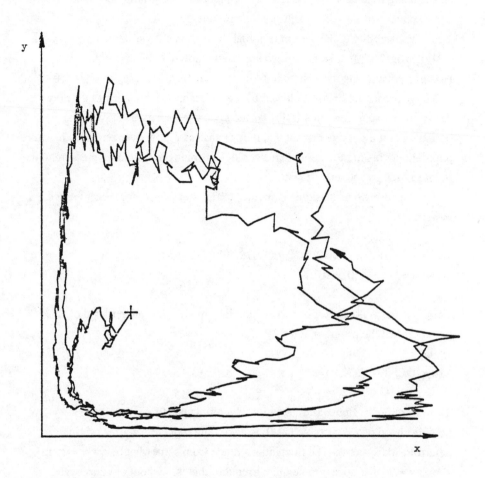

Figure 2.5. A trajectory for a perturbed predator-prey model starting at (1,1). For additional trajectories see [J74].

One can see from Fig. 2.5 that a *trajectory cycle* could be defined as, say, a segment of the trajectory starting and ending on the straight line segment joining (0,0) and (1,1). Actually, in some runs the trajectory might momentarily jump back and forth across the line rather than neatly crossing it, so the idea of a trajectory cycle is not precise. However, experimenting with different runs shows the idea is useful. Particularly long or short (time) cycles are associated with particularly large or small amplitude cycles, respectively. Cycles with large amplitude tend to occur consecutively, as do cycles with small amplitudes.

The stability of the system is probabilistic in that if we start a trajectory at (.01,.01) or (20,20), then the cycles soon settle down to the above pattern. However, displacing the system initially very far from (1,1) say to (100,100), leads to a large predator increase followed by a prey decrease which brings the prey density so close to zero, that is, brings the system trajectory so close to the boundary of the positive orthant that it by chance can jump over zero and "negative populations" result. Presumably the ecological analogue is hunting pressure to the point of local extinction of prey.

Suppose we next ask the computer to find trajectories for the simplified system

$$x(t+\Delta t) = x(t)+[-.1x^2(t)+2.1x(t)]\Delta t+.07x(t)[2RND(1)-1]$$
$$y(t+\Delta t) = y(t) = y(0) = 0$$

This system would correspond to the previous model after the removal of all predators. In the program, line 110 becomes

110   DEF FNF2(X,Y)=0

This simplified system has trajectories which fluctuate in a $x(t)$ value in a neighborhood of $x = 21$ (note $(21,0)$ is a constant trajectory for the original deterministic system). The fluctuations range from somewhat below $x = 15$ to above $x = 30$ and appear to be quite irregular, that is, without obvious cycles.

A *cyclic trajectory* for a difference equation dynamical system is a trajectory that cycles, that is, for which a period (time interval) P exists with $x(t+P) = x(t)$ for all t (here $x(t)$ is the entire state vector). In problem 14 of Chapter 1 such a cyclic trajectory was found with a period of six time units. Actually, a cyclic trajectory for a difference equation system is something of a fluke, a fluke that $x(t+P)$ should be exactly $x(t)$. (Cyclic trajectories for differential equation systems are another matter, as will be explained in Chapter 3.) However, if all trajectories of a probabilistic two-dimensional difference equation dynamical system which start sufficiently close to an annular region in state space tend with time to approach the annular region and cycle within it, we say the system exhibits a *probabilistic limit cycle*.

REFERENCE

[J] C. Jeffries, Probabilistic limit cycles, in *Mathematical Problems in Biology*, ed. P. van den Driessche, Springer, Berlin, 1974, pages 123-131.

APPENDIX

Here is a BASIC program for plotting trajectories starting at (2,1) for the deterministic predator-prey model

$$x(t+\Delta t) = x(t)+[-.1x^2(t)-2x(t)y(t)+2.1x(t)]\Delta t$$
$$y(t+\Delta t) = y(t)+[2x(t)y(t)-.1y^2(t)-1.9y(t)]\Delta t$$

```
100   CLS
110   SCREEN 2
120   GOSUB 300
130   X=2: Y=1: S+.01: ITERATIONS = 100
180   GOSUB 200
190   END
200   'Iterate difference equations
210   GOSUB 350
220   FOR T=1 TO ITERATIONS
230   DX=X*(-0.1*X-2*Y+2.1)*S
240   DY=Y*(2*X-0.1*Y-1.9)*S
250   X=X+DX
260   Y=Y+DY
270   GOSUB 380
280   EXIT
290   RETURN
300   'Draw axes
310   LINE (0,180) - STEP (500,0)
320   LINE (0,0) - STEP (0,180)
330   XF=100: YF=-36: YO = 180
340   RETURN
350   'Plot point
360   PSET (XF*X, YO+YF*Y)
370   RETURN
380   'Draw line to new point
390   LINE - (XF*X, YO+YF*Y)
400   RETURN
```

PROBLEMS

1. Program a computer to evaluate the probabilistic predator-prey model

$$x(t+\Delta t) = x(t)+[-.1x^2(t)-2x(t)y(t)+2.1x(t)]\Delta t+px(t)[2RND(1)-1]$$
$$y(t+\Delta t) = y(t)+[2x(t)y(t)-.1y^2(t)-1.9y(t)]\Delta t+py(t)[2RND(2)-1]$$

with $t = .01$ and perturbation scale constant $p = .07$. List the coordinates of 50 points along the trajectory up to $T1 = 5$ (so about every tenth point). Try different perturbation scale constants including .04, .1, and .2. Comment on the results.

2. Program a computer to evaluate the probabilistic predator-prey system

$$x(t+\Delta t) = x(t)+[-2x^2(t)-.1x(t)y(t)+2.1x(t)]\Delta t+.07x(t)[2RND(1)-1]$$
$$y(t+\Delta t) = y(t)+[2x(t)y(t)-1.9y^2(t)-.1y(t)]\Delta t+.07y(t)[2RND(2)-1]$$

List the coordinates of 50 points along the trajectory up to $T1 = 5$. What would be the ecological interpretation of the relative sizes of the new coefficients in the rate of change functions? What would a graph of a trajectory of this system look like over a long time span?

3. Consider the three component model

$$x_1(t+\Delta t) = x_1(t)+[-x_1^2(t)-x_1(t)x_2(t)+1100x_1(t)]\Delta t$$
$$x_2(t+\Delta t) = x_2(t)+[.1x_1(t)x_2(t)-.9\,x_2(t)^2-x_2(t)x_3(t)\,]\Delta t$$
$$x_3(t+\Delta t) = x_3(t)+[.1x_2(t)x_3(t)-x_3(t)^2]\Delta t$$

Find a constant trajectory in the positive orthant and a constant trajectory with $x_2 = x_3 = 0$, $x_1 \neq 0$, in the nonnegative orthant. Write a BASIC program to find points of the trajectory starting at $(1100,100,10)$ with $\Delta t = .001$.

4. Using the program from problem 3, model the system in which the equation for $x_1(t+\Delta t)$ is changed to

$$x_1(t+\Delta t) = x_1(t)+[-x_1^2(t)-x_1(t)x_2(t)+1100x_1(t)+\cos(\pi t)]\Delta t$$

Use $\Delta t = .001$ and $T1 = 1$. Describe qualitatively how fluctuations in x_1 affect x_3.

5.   Consider the difference equation dynamical system (here we use x_i for $x_i(t)$):

$x_1(t+\Delta t) = x_1(t)+[x_1-x_1x_2]\Delta t$

$x_2(t+\Delta t) = x_2(t)+[2x_1x_2-x_2-x_2^2]\Delta t$

$x_3(t+\Delta t) = x_3(t)+[x_3-x_3x_4]\Delta t$

$x_4(t+\Delta t) = x_4(t)+[2x_3x_4-x_4-x_4^2]\Delta t$

$x_5(t+\Delta t) = x_5(t)+[x_2+x_3-2x_5x_6]\Delta t$

$x_6(t+\Delta t) = x_6(t)+[2x_5x_6-x_6-x_6^2]\Delta t$

$x_7(t+\Delta t) = x_7(t)+[2x_6-x_7x_8-x_7^2]\Delta t$

$x_8(t+\Delta t) = x_8(t)+[2x_7x_8-x_8-x_8^2]\Delta t$

Verify that $(1,1,1,1,1,1,1,1)$ is a constant trajectory for this system. Using a BASIC or other program, $\Delta t = .01$, and an initial state of $(2,1,1,1,1,1,1,1)$, find the state of the system at $t=1,2,3,4,5$. Comment on values of $x_3(t)$ and $x_4(t)$ along the trajectory.

ANSWERS TO PROBLEMS

1. The points of the trajectory for any random number generator should resemble the points in Fig. 2.5. Using $p = .04$ results in not so much an annulus around (1,1) as a disk more or less centered at (1,1). The larger p values result in a few well defined circuits around (1,1) followed by a chance excursion into "negative populations," presumably analogous to the extinction of a species.

2. In this model, auto-regulation dominates the predation interaction. Over a long time span a trajectory for this system more or less fills a disk centered at (1,1).

3. The unique constant trajectory for this model in the positive orthant is (1000,100,10). The unique constant trajectory with $x_2 = x_3 = 0$, $x_1 \neq 0$, is approximately (33.166,0,0).

4. Trajectories for this model generally gravitate to and oscillate irregularly within a small region around (1000,100,10). Fluctuations in x_1 cause fluctuations in x_2 and, indirectly and later, fluctuations in x_3.

5. Typical trajectory values for this model are:

t	x_1	x_2	x_3	x_4	x_5	x_6	x_7	x_8
0	2	1	1	1	1	1	1	1
1	1.052	1.645	1	1	1.195	1.212	1.077	1.036
2	.806	.980	1	1	.942	1.101	1.062	1.137
3	.915	.823	1	1	.938	.947	.965	1.043
4	1.051	.925	1	1	.993	.941	.963	.963
5	1.054	1.052	1	1	1.026	.998	.999	.965

Here $x_3(t)$ and $x_4(t)$ are constants because components 3 and 4 constitute a subsystem which drives the rest of the system and because initially the 3,4 subsystem starts on a constant trajectory, namely $x_3 = x_4 = 1$.

It is, of course, possible to modify the BASIC program for the predator-prey equations by defining additional variables and suitable functions. Another approach is to use a spreadsheet such as Lotus 1-2-3. A listing of cell formulas for a programmed spreadsheet follows:

```
A1:    '{goto}B8~+B6+B8~
A2:    '{calc}
A3:    '/xi(B7-B8)>0~/xgA2~
A4:    '/xq
A6:    'delta t
B6:    0.01
A7:    'max t
B7:    0.99
A8:    'current  t
B8:    +B6+B8
A11:   (F3) +A21
B11:   +B21
C11:   2
A12:   (F3) +A22
B12:   +B22
C12:   1
A13:   (F3) +A23
B13:   +B23
C13:   1
A14:   (F3) +A24
B14:   +B24
C14:   1
A15    (F3) +A25
B15    +B25
C15:   1
A16:   (F3) +A26
B16:   +B26
C16:   1
A17:   (F3) +A27
B17:   +B27
C17:   1
A18:   (F3) +A28
B18    +B28
C17:   1
A18:   (F3) +A28
B18:   +B28
C18:   1
A21:   +A11+(A11-A11*A12 )*$B$6
```

A22: +A12+(2*A11*A12-A12-A12^2)*B6

A23: +A13+(A13-A13*A14)*B6

A24: +A14+(2*A13*A14-A14-A14^2)*B6

A25: +A15+(A12+A13-2*A15 *A16)*B6

A26: +A16+(2*A15*A16-A16-A16^2)*B6

A27: +A17+(2*A16*A17-A18-A17^2)*B6

A28: +A18+(2*A17*A18-A18-A18^2)*B6

Before starting a calculation one must set "Recalculation" to "Rowwise" and assign a macro range name with "Range," "Create," "Name," "A," and range "A1. . A4." The trajectory initial conditions are entered by "Copy" C11. .C18 to A11 . . A18 and entering "CTRL"+"Shift"+A followed by "CTRL"+"Break." The trajectory itself is calculated by entering "Copy" B11. .B18 to A11. .A18 followed by "CTRL"+"Shift"+A.

Lotus 1-2-3 is a very flexible format well suited to experimenting with difference equation dynamical systems of dimension 5 to 500.

CHAPTER THREE
FORMALIZING THE NOTION OF STABILITY

3.1 The Concept of Ecosystem Stability

Suppose a train proceeds down a track toward its destination. Suppose the remaining distance to its destination is always decreasing by at least 100 km/hr so long as the train is at least 10 km from its destination. Suppose the initial distance from the train to the destination is 510 km. The reader can no doubt appreciate that the train can take at most five hours before lying within 10 km of its destination. This simple idea could readily be extended to other dimensions, say the three-dimensional rendezvous of a spacecraft with a manned satellite. The details of motion are unknown and unneeded; yet the conclusion is obvious.

Next suppose the distance to destination is not always decreasing, or at least not known to be always decreasing in a simple way. Suppose instead that "distance to destination" is decreasing according to some observer viewing the progress of the vehicle through some sort of imperfect lens. The imperfect lens makes straight lines wavey lines, but does not distort overly much the concepts of near and far. It seems reasonable to assume that upper limits on time to a neighborhood of the destination could still be calculated, perhaps depending upon the definition of "overly much."

Such is the essential idea of Lyapunov's theory of stability of motion: details of motion are not essential to guarantee the approach of a destination so long as some measure of remaining distance is decreasing. The point of this chapter is to make that idea mathematically precise.

In this book we shall assume that the concept "ecosystem stability" has two parts. First it means that there exists an attractor trajectory (or attractor region as in Chapter Seven) for a dynamical system model of the ecosystem. Actually, the very existence of an ecosystem might be interpreted as the existence of an attractor trajectory or attractor region for any properly designed model. Thus, in discussing stability of an ecosystem we must delineate the size of the *basin of attraction* of the attractor trajectory or region. By basin of attraction we mean the set of points in state space-time thought of as initial values of trajectories which asymptotically approach the attractor trajectory or which enter the attractor region after a finite time interval. Certainly more than one attractor trajectory or attractor region might exist; in that case, a description of how state space-time is partitioned into basins of attraction is needed.

37

First of all we need to specify how a trajectory of an ecosystem model can be efficiently computed, that is, how the train goes down the track.

3.2 The Relation Between Difference and Differential Equations

As everyone who has studied calculus knows, the derivative of a function is approximated by the slope of the hypotenuse of a little triangle, in fact, an arbitrarily small triangle. It seems that the concept that either makes or breaks the calculus student is this: even though the sides of the triangle can become arbitrarily small, the ratio of the length of the vertical side to the length of the horizontal side generally becomes, for a differentiable function, a particular number, the rate of change of the function. The triangle we have in mind is shown in Fig. 3.1.

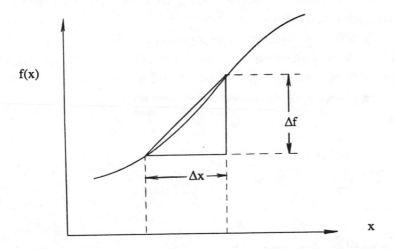

Figure 3.1. The fundamental triangle of differential calculus.

Does the reader believe it? Can you imagine a triangle shrinking so that the ratio of lengths of its sides approaches a fixed number, even though the lengths themselves approach zero? If so, then you grasps an essential idea; otherwise, you need to think harder. Perhaps it helps to imagine first the special case of a triangle shrinking so that each version is similar to the original, so the ratios of the sides are constants. Generally in computing derivatives of (differentiable) functions, the triangle changes shape slightly and ever more slowly as it shrinks.

The symbolisms

$$df/dx = \lim_{\Delta x \to 0} \frac{\Delta f}{\Delta x}$$

and

$$\dot{x}(t) = dx/dt = \lim_{\Delta t \to 0} \frac{\Delta x}{\Delta t}$$

are doubtless familiar. Note that the dot notation $\dot{x}(t)$ always refers to differentiation with respect to time t.

The reader should certainly be able to affirm that if $x(t)$ happens to be t^2+t+2, then $\dot{x}(t) = 2t+1$. If $t = 8$, then $x(8) = 74$ and $\dot{x}(8) = 17$. If the derivative of a function exists (so the function is called differentiable), then it is itself a function. The results from calculus used in this book can be found in [F] or any other first-rate multivariate calculus text.

The difference $x(t+\Delta t)-x(t)$ is the change in x value from t to $t+\Delta t$. Shorthand for this difference is $\Delta x(t)$. Thus associated with the <u>difference</u> equation dynamical system $x_i(t+\Delta t) = x_i(t)+f_i(x,t)\Delta t$ are the *rate of change functions* $f_i(x,t)$.

The analogous *differential equation dynamical system* is

$$\dot{x}_i = f_i (x,t)$$

or, as n-vectors,

$$\dot{x} = f (x,t)$$

In differential equation dynamical systems, the functions in $f(x,t)$ are called the *system functions*. An n-vector of functions $x(t)$ satisfying the equation is called a *trajectory of the dynamical system*.

Of course, if we start with a difference equation dynamical system, there is always a way--namely, brute force--to determine trajectories of the system. Differential equation dynamical systems are more subtle. Let us refer to a central theorem on the existence of trajectories of differential equation dynamical systems.

Trajectory Existence Theorem. Suppose the system functions f(x,t) for the dynamical system ẋ(t) = f(x,t) are continuous and have continuous derivatives in some open set of space-time. Then to any initial state in that open set is associated a local unique trajectory x(t) solving ẋ(t) = f(x,t).

(The definitions of calculus terminology used in the statement of this theorem can be found in the Appendix.)

The Trajectory Existence Theorem is a standard result in dynamical system theory and can be studied in, for example, [S, p. 8]. For the sake of mathematical convenience we shall assume in this book that all unspecified system functions meet the conditions of this theorem for all time and throughout the positive orthant or throughout state space, depending on context.

With the increasing availability of computers and use of computers by nonmathematicians, it seems to be ever easier to be successful at just about anything except pure mathematics while believing in the existence of only a finite number of numbers. If one uses only the decimal numbers with, say, 99 digits to the left and 99 digits to the right of the decimal, together with the additive inverses (negatives) of such numbers, one can do much science and engineering with a computer . It is logical to question, then, the necessity of studying the <u>differential</u> as opposed to the conceptually simpler <u>difference</u> equation viewpoint.

Studying differential equations is necessary in modeling because the generic essence of a model is often revealed only in the limit $\Delta t \to 0$, not with numerous experimental choices of "small" Δt. Qualitative knowledge based on mathematical ideas of how systems work is without a doubt a modeling tool as useful as computer simulations.

Let's consider an example of such usefulness. Consider the two-dimensional system

$$\dot{x}_1 = x_2$$
$$\dot{x}_2 = -x_1$$

The two system functions are here $f_1(x_1, x_2, t) = x_2$ and $f_2(x_1, x_2, t) = -x_1$. (These system functions obviously fulfil the conditions in the Trajectory Existence Theorem.) The reader should verify by differentiation that

$$x_1(t) = A \cos(t) + B \sin(t)$$
$$x_2(t) = -A \sin(t) + B \cos(t)$$

is a trajectory for the system where A and B are arbitrary constants. Clearly A and B are determined by initial conditions. If the initial time is t = 0, then $x_1(0) = A$ and $x_2(0) = B$. If A = B = 0, then we have a constant trajectory at (0,0). Otherwise, by evaluating $x_1(t)^2 + x_2(t)^2$ the reader should be abel to verify that the trajectory traces out a circle of radius $(A^2 + B^2)^{1/2}$ centered at (0,0) and passes through its initial poing every 2π time units. (Derivatives of cos(t) and sin(t) are listed in the Appendix.) A typical trajectory is shown in Fig. 3.2.

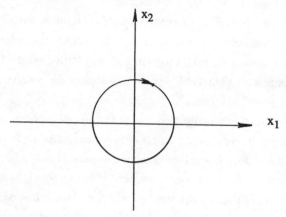

Figure 3.2. The trajectory (cos(t)+2sin(t), -sin(t)+2cos(t)) starting at (1,2).

Trajectories of the difference equation analogue of the system would not, of course, be points exactly on a circle and would generally never pass through a given point more than once. In fact, points of such trajectories would lie on curves which would spiral away from (0,0), regardless of how small Δt were chosen. Over the long run, there is a big difference between a trajectory which goes around and around a circle and a trajectory which spirals outward! This example warns us of potential modeling subtleties.

Fortunately "stability" eliminates such bothersome possibilities. Roughly speaking, if a model has this property, stability, then whenever Δt is "suitably small" or smaller, computer simulations do give an accurate and consistent qualitative picture of system dynamics.

Almost as fortunately, there are "better" ways to go down the track, that is, to go from differental equation dynamical systems to difference equation dynamical systems. Before we leave difference equation theory, let us consider such a better way.

The way we have been using, the way based on the triangle of calculus, is

called the *forward-difference procedure* of numerical integration; it is also
known as *Euler's method*. It takes

$$\dot{x}_i(t) = f_i(x,t)$$

and gives us

$$x_i(t+\Delta t) = x_i(t) + f_i(x(t),t)\Delta t$$

Of course, it might be logical in a modeling effort to <u>start</u> with a difference equation
dynamical system. If we try to model insect populations from year to year based on
levels at July 1 of each year, then for better or worse we inherently have a
difference equation model and are stuck with $\Delta t = 1$ year.

     If the model is based on differential equation thinking and we wish to convert
the model into difference equations for the purpose of running computer
simulations, then we enter a realm of choices of numerical integration techniques.
A technique which compared to the forward-difference procedure often gives better
numerical accuracy using fewer calculations (larger Δt) is the *double-approximation
procedure* or *Runge-Kutta method of order 2*, namely,

$$x_i(t+\Delta t) = x_i(t) + .5[f_i(x(t),t) + f_i(x(t)+f(x(t),t)\Delta t,t+\Delta t)]\Delta t$$

As Δt becomes arbitrarily small, this version of $x_i(t+\Delta t)$ becomes arbitrarily close to
the forward-difference version. The difference is, again, that in practice larger Δt
can generally be used with the double-approximation procedure to achieve the same
or better accuracy. Of course, larger Δt over a given time interval means fewer
computation steps.

     It is possible to use algebra and geometry to show that the ***double-
approximation procedure*** is equivalent to the following sequence:

    1. use the <u>forward-difference</u> procedure to find the point
    $p = x(t)+f(x(t),t)\Delta t$;
    2. use the <u>forward-difference</u> procedure starting at p to find
    the point $q = p+f(p,t+\Delta t)\Delta t$;
    3. set $x(t+\Delta t) = .5 \, x(t) + .5 \, q.$

The double-approximation procedure is explored in problem 3.4.

Numerical integration is a body of mathematical knowledge in itself. A general reference is [Mar].

3.3 Limit Cycles

An important feature of attractor trajectories generally is the possibility of time dependence. A nonconstant attractor trajectory serves as a sort of moving target pursued by the model, as opposed to a constant attractor trajectory, a stationary target.

In some models, including some without time dependence in their system functions, limit cycles occur. A *limit cycle* is a simple closed loop in state space with two properties. First, any trajectory starting at any time at any point on the limit cycle stays on the limit cycle forever and is a cyclic trajectory with period T, T being a characteristic of the limit cycle. Second, any trajectory starting sufficiently close to the limit cycle must asymptotically approach some trajectory in the limit cycle.

Although much effort has gone into the study of limit cycles in ecosystem models (see [May] for a good introduction), I mention them reluctantly in this book. Every modeler should know about limit cycles as a possible mathematical phenomenon. But outside the laboratory, the roughly cyclic nature of some natural populations cannot be a simple expression of a limit cycle. If it were, then somehow starting the natural system at the same state but at two different times would lead to two system trajectories which would subsequently always remain separated by at least some fixed minimal amount. As a mathmatician I feel that while there are likely plenty of attractor trajectories which happen to be roughly cyclic (sometimes with cycles in cycles), the existence of a true natural ecosystem limit cycle is unlikely.

The following abstract dynamical system has a limit cycle:

$$\dot{x}_1 = (1-x_1^2-x_2^2)x_1-x_2$$
$$\dot{x}_2 = (1-x_1^2-x_2^2)x_2+x_1$$

Clearly $\hat{x}(t) = (0,0)$ is a constant trajectory for this system. Some other trajectories have $x(t) = (x_1(0)\cos(t)-x_2(0)\sin(t), x_1(0)\sin(t)+x_2(0)\cos(t))$ where $x_1(0)$ and $x_2(0)$ are any two numbers the sum of the squares of which is one (see problem 3.6). The latter type of trajectory traces out the unit circle (circle of radius one) every $T = 2\pi$ time units. In fact, the unit circle is a limit cycle for this system; typical trajectories,

43

as derived in problem 3.6, are shown in Fig. 5.1.

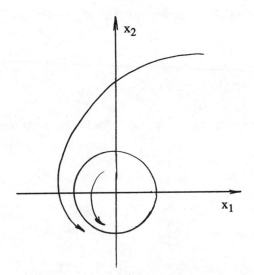

Figure 3.3. A two-dimensional dynamical sytem with a limit cycle.

The key feature in Fig. 3.3 is that trajectories other than (0,0) and those which start on the unit circle asymptotically approach some cyclic trajectory which lies in the unit circle. Different initial values generally result in asymptotic approach of different cyclic trajectories (moving targets). Note that distinct trajectories starting on the unit circle remain forever separated by a fixed distance; thus there is no attractor trajectory for the system.

There are many references in the literature to limit cycles as potential paradigms of ecosystem behavior (the account of Nicholson's blowfly experiments in a laboratory in [May] is a classic). But rather than limit cycles we shall emphasize in this book time-driven models with attractor trajectories which might, given certain simplifying assumptions, happen to be cyclic.

3.4 Lyapunov Theory

Recall that the distance $d(x,y)$ between two points x and y in n-dimensional state space is defined to be

$$d(x,y) = \left[\sum_{i=1}^{n} (x_i - y_i)^2 \right]^{1/2}$$

This distance is not always the "best" way to measure how close a state x is to the current value of an attractor trajectory x(t). Instead, "closeness" can be measured by a Lyapunov function $\Lambda(x,t)$, as we now illustrate.

A trajectory starting at (2,1) for the hypothetical and simplistic predator-prey model

$$\dot{x}_1 = x_1 (-.1x_1 - 2x_2 + 2.1)$$
$$\dot{x}_2 = x_2 (2x_1 - .1x_2 - 1.9)$$

is shown in Fig. 3.4. (The sketch may be obtained numerically as in Chapter Two.)

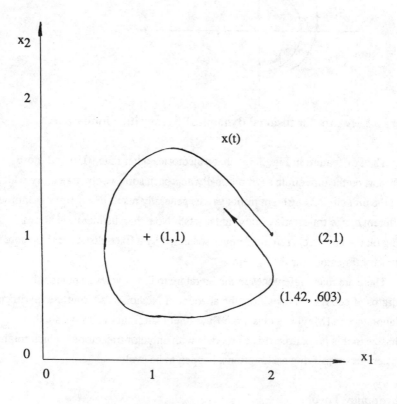

Figure 3.4. A trajectory of a two-dimensional model.

Although the trajectory in Fig. 3.4 seems to approach asymptotically the constant trajectory $(1,1)$, we note that at certain points the distance between the trajectory and $(1,1)$ is apparently increasing. Using calculus the rate of change of distance between an arbitrary trajectory $\hat{x}(t)$ and the constant trajectory $\hat{x}(t) = (1,1)$ is

$$d(x(t),\hat{x}(t)) = -\frac{1}{2} [(x_1-1)^2+(x_2-1)^2]^{-1/2} [2(x_1-1)\dot{x}_1+2(x_2-1)\dot{x}_2]$$

$$= -[d(x,x)]^{-1/2} [(x_1-1)x_1(-.1x_1-2x_2+2.1) + (x_2-1)x_2(2x_1-.1x_2-1.9)]$$

The trajectory starting at $(2,1)$ passes close to $(1.24, .603)$ whereupon the above rate of change is about $+.412$. Thus there are indeed trajectories and times for which distance to $(1,1)$ is actually increasing.

However, let us consider another way to measure the current state of the system relative to $\hat{x}(t) = (1,1)$. Let the function $\Lambda(x_1,x_2)$ be defined by

$$\Lambda(x_1,x_2) = x_1 - \ln(x_1) - 1 + x_2 - \ln(x_2) - 1$$

(In the Cyrillic alphabet, Λ is the first letter of "Lyapunov;" Λ is pronounced "ell.")

Along any trajectory $x(t)$ in the positive orthant the rate of change of Λ is

$$\dot{\Lambda} = \dot{x}_1 - x_1^{-1}\dot{x}_1 + \dot{x}_2 - x_2^{-1}\dot{x}_2$$

$$= (x_1-1)x_1^{-1}\dot{x}_1 + (x_2-1)x_2^{-1}\dot{x}_2$$

$$= (x_1-1)[-.1(x_1-1) - 2(x_2-1)] + (x_2-1)[-.1(x_1-1) + 2(x_2-1)]$$

$$= -.1(x_1-1)^2 - 2(x_1-1)(x_2-1) - .1(x_2-1)^2 + 2(x_2-1)(x_1-1)$$

$$= -.1(x_1-1)^2 - .1(x_2-1)^2$$

Thus Λ always decreases along any trajectory in the positive orthant with one exception: along the constant trajectory $\hat{x}(t) = (1,1)$ Λ is a constant, zero.

Suppose we have an analytic function defined on some regions of n-dimensional space which has nonnegative values; Λ is, of course, such a function. A *level set* is the set of all points in space for which the value of a given function is a certain nonnegative number λ. Because we want to deal with realistic, time dependent dynamical systems and attractor trajectories, we need to consider level sets in state space-time. Thus we extend the definition of $\Lambda(x)$ to $\Lambda(x,t) = \Lambda(x)$; Λ thereby becomes a function in state space-time, not just state space.

46

Some representative level sets associated with $\Lambda(x,t)$ and the function $d(x,\hat{x}(t))$ are shown in Fig.3.5. Of course, the dynamical system represented in Fig. 3.5 is actually autonomous (time independent), as are $\Lambda(x,t)$ and the attractor trajectory, $\hat{x}(t)$. Thus the model is perhaps unduly simple. We shall, however, deal with time dependent models in the Chapter Six.

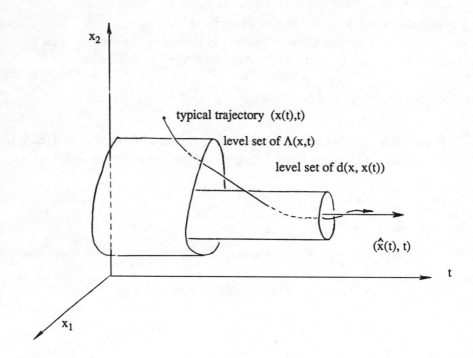

3.5. Relations of level sets, a constant attractor trajectory $\hat{x}(t)$, and a typical trajectory $x(t)$.

The above level sets suggest the concepts of "inside" and "outside" relative to a trajectory $\hat{x}(t)$. Suppose there exist two level sets L_1 and L_2 in state space-time with the property that L_1, L_2 and $\hat{x}(t)$ have no points in common. L_1 is *inside* L_2 if each ray beginning at every point $\hat{x}(t)$ and perpendicular to the time axis intersects first L_1 and then L_2. Thus in Fig.3.5 the level set $L = \{(x,t) \mid d(x,\hat{x}(t)) = \varepsilon\}$ is inside the level set $L = \{(x,t) \mid \Lambda(x,t) = \lambda\}$.

If $L_1, L_2, \ldots, L_q$ are all level sets about $\hat{x}(t)$ with L_i inside L_{i+1} for each $i = 1, 2, \ldots, q - 1$, then we say the level sets are *nested* about $\hat{x}(t)$. If L_1 and L_2 are level sets and L_1 is inside L_2 relative to $\hat{x}(t)$, then we also say L_2 is *outside* L_1. Likewise a point (x,t) not in a level set L can be inside or outside L, depending on whether a ray perpendicular to the time axis and starting at $\hat{x}(t)$ reaches the point (x,t) or L first.

Let R denote a region of state space-time, possibly the positive orthant and all of time. We shall be concerned with R at least big enough to include all level sets $\Lambda = \lambda$ of Λ for $0 \leq \lambda \leq \delta$ for some positive number δ. We shall also use the level set of (positive) radius ε about $\hat{x}(t)$, meaning for each time t the set of all points in state space at a distance ε from $\hat{x}(t)$.

We are now equipped to prove a version of M. A. Lyapunov's important result.

Lyapunov Theorem. Suppose we have a dynamical system $\dot{x} = f(x,t)$ in a region R of n–dimensional space with trajectory $\hat{x}(t)$ and an analytic, nonnegative function $\Lambda(x,t)$, all related as follows:

1. **for any level set L of contained in R, there exist level cylinders inside and outside L.**

2. **for any level cylinder C, there exists a level set L_1 of Λ inside C; if the radius of C is smaller than a number determined by**

   $\hat{x}(t)$, **then C is inside some set L of Λ in R;**

3. **given a level cylinder in R, there is a positive number β such that $\dot{\Lambda} \leq -\beta$ at all points in R and outside the level cylinder.**

If these three conditions are met, then $\hat{x}(t)$ is an attractor trajectory for the dynamical system.

48

Proof. First we show that $\hat{x}(t)$ is asymptotically approached by all nearby trajectories. Let ε be any positive number and let $x(t_0)$ in R be the initial state of an arbitrary trajectory. Suppose $x(t_0)$ is not already inside the level cylinder C of radius ε. Our task is to show that x(t) will be inside C within a finite time interval determined by distance between $x(t_0)$ and $\hat{x}(t_0)$.

Let L denote the level set of Λ which contains $x(t_0)$. If C is not inside L, choose a smaller value for ε so it is. Using conditions 1 and 2 we can choose a level set L_1 of Λ and level cylinders C_1 and C_0 such that C_1, L_1, C, L, C_0 are nested level sets (in the given order) about $\hat{x}(t)$. This arrangement of level sets is shown in Fig. 3.6.

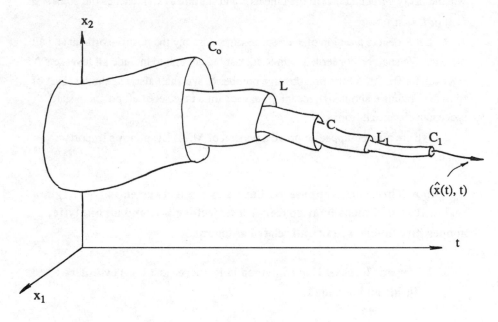

Figure 3.6. Level sets and level cylinders as needed in the proof of the Lyapunov Theorem. Only two of n spatial dimensions can be shown.

By condition 3, $\dot{\Lambda}$ is less than or equal to some -β outside C_1. Let λ be the value of Λ on L and let $λ_1$ be the value of Λ on L_1. It follows that the arbitrary trajectory can take at most $(λ-λ_1)/β$ time units to enter L_1. Since λ can be bounded in terms of the radius of C_0 and since that radius is greater than the distance from $x(t_0)$ to $\hat{x}(t_0)$, the time interval taken before the trajectory must enter C can be bounded by the initial distance from $x(t_0)$ to $\hat{x}(t_0)$.

Let C be a cylinder of arbitrarily small radius about $\hat{x}(t)$. Let L_1 be a level set of Λ and let C_1 be a level cylinder so that C_1, L_1, and C are nested about $\hat{x}(t)$. No trajectory starting in C_1 can ever escape L_1 because $\dot{\Lambda}$ is negative on L_1; all trajectories passing through L must puncture L inwardly. Thus any trajectory starting closer to $\hat{x}(t)$ than the radius of C_1 must thereafter remain within the radius of C of $\hat{x}(t)$.

The above observations imply that $\hat{x}(t)$ is an attractor trajectory.

In the case of the above two–dimensional predator–prey model, $\Lambda = -.1(x_1-1)^2 -.1(x_2-1)^2$. Thus outside the level cylinder of radius ε, $\dot{\Lambda} \le .1\varepsilon^2$. To satisfy condition 3 we need only set $\beta = .1\varepsilon^2$. Note also that R can be taken to be the entire positive orthant and all time.

3.5   The Trapping of Trajectories

Recall that the *positive orthant* of n-dimensional state space refers to all points $x = (x_1, x_2, ..., x_n)$ with each $x_i > 0$. The *nonnegative orthant* refers to points with $x_i \ge 0$. We shall have need of the concept *finite boundary of the positive orthant*, meaning all points in the nonnegative orthant with finite coordinate values and at least one coordinate value zero.

In population dynamics and many other applications of modeling, state space lies in the nonnegative orthant. For a given type of model, how can we be certain that embarrassing negative populations do not arise? In other words, we want to be sure that trajectories that start in the positive orthant are trapped therein and cannot leave, at least through the finite boundary of the positive orthant.

Suppose the system functions of a dynamical system can be written (not uniquely) as a sum

$$\dot{x}_i = x_i g_i(x,t) + h_i(x,t)$$

where each $h_i(x,t)$ is nonnegative. Not every dynamical system has system functions which can be written as such a sum, of course, but every ecosystem model does because energy can only be removed from a compartment (some $g_i(x,t)<0$) if the amount of energy in the compartment is positive. Energy can be added to an empty compartment through detritus donation, but this can be lumped in the h part of the system functions.

The following is a simple theorem which has extensive fortunate implications for ecosystem modeling.

Trajectory Trapping Theorem. Suppose a dynamical system is given by

$$\dot{x}_i = x_i g_i(x,t) + h_i(x,t)$$

where: each $g_i(x,t)$ and each $h_i(x,t)$ is defined and continuous for all x in the nonnegative orthant and all t, and each $h_i(x,t) \geq 0$. Then no trajectory for the system which starts in the positive orthant reaches the finite boundary of the positive orthant after a finite time interval.

Proof. Suppose such a trajectory exists. Without loss of generality we can assume the trajectory starts at t_0 in the positive orthant, stays in the positive orthant for all $t_0 < t < t_1$, and finally reaches the finite boundary of the positive orthant at $t = t_1$. Furthermore, without loss of generality, we can assume that $x_1(t_1) = 0$. Consider the function defined for $t_0 \leq t < t_1$ by $y(t) = \ln(x_1(t))$; it follows that $y(t)$ is

continuous for $t_0 \leq t < t_1$ and $\lim_{t \to t_1} y(t) = -\infty$. Now $\dot{y}(t) = [x_1(t)]^{-1} \dot{x}_1(t) \geq g_1(x(t),t)$.

Since g_1 is continuous in the nonnegative orthant, $g_1 \geq -P$ for some positive number P along the complete trajectory, $t_0 \leq t \leq t_1$. Thus
$y(t) = y(t) - y(t_0) + y(t_0) \geq -P(t-t_0) + y(t_0)$
Thus $\lim_{t \to t_1} y(t) \geq -P(t_1 - t_0) + y(t_0) > -\infty$; this contradicts our previous calculation of
$\lim_{t \to t_1} y(t)$ and so shows no such trajectory can exist.

REFERENCES

[F] Wendell H. Fleming, *Functions of Several Variables* , Springer, Berlin-New York, 1977.

[Mar] M. J. Maron, *Numerical Analysis: A Practical Approach*, Macmillan, New York, 1987.

[May] R. M. May, *Stability and Complexity in Model Ecosystems*, Princeton U. Press, 1973.

[S] D. Sánchez, *Ordinary Differential Equations and Stability Theory* , Freeman,

San Francisco, 1968.

[W] J. L. Willems, *Stability Theory Dynamical Systems,* Nelson, London, 1970.

APPENDIX

This appendix is a reference list of calculus facts used in modeling theory.

Numbers. By the term "numbers" in this book, we usually mean points on the real number line. When dealing with general roots (eigenvalues) of the characteristic polynomial, complex numbers are needed, numbers of the form a+bι where ι is a square root of -1.

Functions. A function f is a machine; if we give it a number x, it gives us a number f(x). Some functions consume points in n-dimensional space and give us a number. In n-dimensional space a function which can be written as a (finite) sum of products of positive integer powers of components of x is called a *multinomial*. In the special case of n = 1, a multinomial is called a *polynomial*.

Limits. A limit process is rather like the concept of the end of a piece of string; the end as an entity is not the same as any point on the string, yet without the string there is no end.

    If f(x) is a real function defined for all real numbers x, then the *limit of f(x) as x approaches x_0*, if the limit exists, is written

$$\lim_{x \to x_0} f(x)$$

The limit does exist precisely if there is a unique number L with the following property: for any number $\varepsilon > 0$ there exists a number $\delta > 0$ (δ determined by ε) such that $|x-x_0| < \delta$ implies $|f(x) - L| < \varepsilon$. In such a case we write

$$\lim_{x \to x_0} f(x) = L$$

If $L = f(x_0)$, then f is called *continuous at x_0*. If such limits exist regardless of choice of x_0, then f is just called *continuous*. Roughly speaking, any function which can in principle or in practice be sketched in (x,f(x))-space without lifting one's pencil from one's paper is continuous. If f(x) is a real-valued function defined on n-dimensional space, then f is called continuous if for any n-1 numbers $a_1, a_2,...,a_n$ (with a_i missing), the function $f(a_1,a_2,...,x,...,a_n)$ is continuous in the one-dimensional sense. Any multinomial is continuous throughout n-dimensional space. Any fraction of multinomials is continuous throughout that portion of n-dimensional space in which the denominator is nonzero.

Slopes. The slope of the straight line in the (x,y)-plane joining the points (x_1,y_1) and (x_2,y_2) is, of course, $(y_2-y_1)/x_2-x_1)$. The slope of the function f(x) between x_1 and x_2 is the slope of the hypotenuse of the triangle shown in Fig. 3.1. If f(x) is a function defined for x in n-dimensional space, then in direction i, i = 1,2,...,n, the slope of f between point $x_0 = (x_{01},x_{02},...,x_{0n})$ and $(x_{01},x_{02},...,x_{0i}+ \varepsilon,...,x_{0n})$ is

$$\frac{f(x_{01},x_{02},...,x_{0i} + \varepsilon,...,x_{0n})-f(x_0)}{\varepsilon}$$

Differentiation. If the limit as $\varepsilon \to 0$ of the above slope exists for all i = 1,2,...,n then f is called differentiable at x_0. Each such limit is labeled $(\partial f/\partial x_i)(x_0)$ and is called the *partial derivative of f at x_0 with respect to x_i*. Let us assume in the remainder of this Appendix that all derivative limits exist throughout the positive orthant.

With one variable we call the only partial derivative available just the derivative and denote it as (df/dx)(x) or just df/dx.

The derivative of a function of one variable is itself a function of one variable. Here is a list of familiar functions and their derivatives.

function	derivative
x	1
cx(c is a constant)	c
c	0
x^n (n is a positive integer)	nx^{n-1}
x^n (n is a nonzero number, x>0)	nx^{n-1}
ln(x) (x>0)	$1/x$
$e^x = \exp(x)$	$e^x = \exp(x)$
sin(x)	cos(x)
cos(x)	-sin(x)

If f(x) and g(x) are differentiable functions, then we have the following additional rules.

function	derivative
f(x)+g(x)	df/dx+dg/dx
f(x) g(x)	(df/dx) g(x)+f(x) (dg/dx)
f[(g(x)]	df/dx[g(x)] (dg/dx) (*Chain Rule*)

Analytic Functions. A function is called *analytic* over a subset of n-dimensional space if it can be expressed as a multinomial or as an infinite multinomial with finite values over the subset. If a function is analytic, then the terms in its expression as a multinomial or infinite multinomial can be algebraically rearranged into a special form called its *Taylor series* [F, p. 49].

Sets. In n-dimensional space an open ball is a set of points satisfying $d(x_1 x_0) < \varepsilon$, that is, the points closer than ε (the radius of the ball) to x_0 (the center of the ball). Any union of a finite number or infinite number of open balls is called an *open set*. For example, the positive orthant of n-dimensional space is an open set. The complement of an open set is a *closed set* [F, p. 164].

PROBLEMS

Most of the following problems deal with abstract mathematical models, not models designed along the lines of simplified ecosystems. For better or worse, one needs to acquire a certain degree of Lyapunov literacy before being able to comment meaningfully on Lyapunov stability of more complicated systems.

1. Convert the difference equation

$$x_1(t+\Delta t) = x_1(t)+[-.1x_1(t)] \Delta t$$

into differential equation by finding the limit of the difference equation system as Δt approaches zero.

2. Convert the difference equations

$$x_1(t+ \Delta t) = x_1(t)+[-x_1(t)+x_2(t)] \Delta t$$
$$x_2(t+ \Delta t) = x_2(t)+[-x_1(t)-x_2(t)] \Delta t$$

into differential equations by finding the limit of the difference equation system as Δt approaches zero.

3. Convert the following difference equations into differential equation form

$$x_1(t+ \Delta t) = x_1(t)+[2x_1(t)\cos(x_1(t))] \Delta t$$
$$x_2(t+ \Delta t) = x_2(t)+[2] \Delta t$$
$$x_3(t+ \Delta t) = x_3(t)+[x_1(t)x_2(t) \, x_3(t)] \Delta t$$

4. In the first section of this chapter we considered the two-dimensional dynamical system

$$\dot{x}_1 = x_2$$
$$\dot{x}_2 = -x_1$$

Write a program in BASIC or another language which uses the forward-difference procedure to calculate the value at $t = 1$ of the trajectory of this system which starts at (1,1). Use $\Delta t = .01$ for one run and $\Delta t = .001$ for another. Then repeat everything with the double-approximation procedure. Along a trajectory of the differential system, the distance from the trajectory value to the origin should be constant, namely $2^{1/2} \cong 1.414213562$. Compare with $2^{1/2}$ the four distances resulting from the four simulations.

5. Consider the abstract mathematical model

$$\dot{x}_1 = -x_1 - x_2$$
$$\dot{x}_2 = +x_1 - x_2$$

Show $x(t) = (0,0)$ is the only constant trajectory for this model. Compute the rate of change of $\Lambda(x_1,x_2,t) = (x_1{}^2+x_2{}^2)/2$ along an arbitrary trajectory. Describe the level sets of Λ. Given a level cylinder of radius ε in state space-time, find β in terms of ε so that $\Lambda < -\beta$ outside the level cylinder.

6. Show $x(t) = (x_1(0)\cos(t) - x_2(0)\sin(t), x_1(0)\sin(t) + x_2(0)\cos(t))$ is a trajectory with $x_1(t)^2+x_2(t)^2 = 1$ for the system

$$\dot{x}_1 = (1-x_1^2-x_2^2)x_1 - x_2$$
$$\dot{x}_2 = (1-x_1^2-x_2^2)x_2 + x_1$$

where $x(0) = (x_1(0),x_2(0))$ is on the unit circle, that is, $x_1^2(0) + x_2^2(0) = 1$.

7. In two dimensional space describe the level sets of $\Lambda(x_1,x_2) = (x_1^2+x_2^2-1)^2$ with values 0, .25, 1, 4, and 10^6.

8. Show that the derivative of $\Lambda(x_1,x_2,t) = (x_1^2+x_2^2-1)^2$ along an arbitrary trajectory of

$$\dot{x}_1 = (1-x_1^2 -x_2^2)x_1 -x_2$$
$$\dot{x}_2 = (1-x_1^2 -x_2^2)x_2 +x_1$$

is

$$\dot{\Lambda} = -4(x_1{}^2+x_2{}^2-1)(x_1{}^2+x_2{}^2)$$

For what values of (x_1,x_2,t) is $\dot\Lambda = 0$? What is the sign of $\dot\Lambda$ elsewhere?

9. Consider the abstract dynamical system defined over all of two-dimensional state space:

$$\dot{x}_1 = -x_1 + 3x_2$$

$$\dot{x}_2 = -x_1 - x_2$$

Show $\hat{x}(t) = (0,0)$ is the only constant trajectory for this system. Define $\Lambda(x,t) = x_1^2 + k\,x_2^2$ where k is a constant. Find a positive value for k so that $\dot\Lambda < 0$ along all nonconstant trajectories of the system. Using such k sketch typical level sets of Λ. Verify that conditions 1 and 2 of the Lyapunov Theorem are met. Given a level cylinder in state space-time about x(t) with radius ε, find a number β so that $\dot\Lambda < -\beta$ outside the cylinder, that is, verify condition 3 of the Lyapunov Theorem for this model. Show, therefore, that $\hat{x}(t)$ is an attractor trajectory for the system and that the basin of attraction R is all of state space-time.

10. Consider the abstract dynamical system

$$\dot{x}_1 = -x_1 + 3r(x,t)x_2$$
$$\dot{x}_2 = -r(x,t)x_1 - x_2$$

where r(x,t) is an arbitrary analytic functions defined throughout state space-time. Using information from the previous problem, show the constant trajectory $\hat{x}(t) = (0,0)$ is an attractory trajectory for this system, the basin of attraction being all of state space-time.

11. Consider the abstract dynamical system

$$\dot{x}_1 = -a_1(x,t)x_1 = 3r(x,t)x_2$$
$$\dot{x}_2 = -r(x,t)x_1 - a_2(x,t)x_2$$

where $r(x,t)$, $a_1(x,t)$, and $a_2(x,t)$ are analytic functions defined throughout state space-time. Suppose $a_1(x,t)$ and $a_2(x,t)$ have values greater than one, but otherwise let $a_1(x,t)$, $a_2(x,t)$, and $r(x,t)$ be arbitrary. Using information from problem 6, show the constant trajectory $\hat{x}(t) = (0,0)$ is an attractor trajectory for this system, the basin of attraction being all of state space-time.

12. Consider the model of a simplified ecosystem with three compartments given by

$$\dot{x}_1 = x_1(-x_1-x_2+2)$$
$$\dot{x}_2 = x_2(2x_1-x_2-x_3)$$
$$\dot{x}_3 = x_3(x_2-x_3)$$

Here compartment 3 preys upon compartment 2, and compartment 2 preys upon compartment 1, using the terminology of Chapter Five. Show $\hat{x}(t) = (1,1,1)$ is a constant trajectory for this system. Let $\Lambda(x_1,x_2,x_3,t)$ be defined by

$$\Lambda = 2(x_1-\ln(x_1)-1) + (x_2-\ln(x_2)-1) + (x_3-\ln(x_3)-1)$$

Show the rate of change $\dot{\Lambda}$ of Λ along an arbitrary trajectory in the positive orthant is

$$\dot{\Lambda} = -2(x_1-1)^2 - (x_2-1)^2 - (x_3-1)^2.$$

13. In (x,y)-space sketch the function $y = x - \ln(x) - 1$ (for $x > 0$) showing its minimum, its vertical asymptote, and its relationship with the linear function $y = x - 1$.

14. In n–dimensional space consider the function $\Lambda(x) = \sum_{i=1}^{n} x_i - \ln(x_i) - 1$. Denote a ray as $(1+sa_1, 1+sa_2,\ldots,1+sa_n)$ where $(1,1,\ldots,1)$ is the endpoint of the ray, s is a nonnegative variable, and $a_1, a_2,\ldots, a_n$ are arbitrary constants, not all zero. Show Λ increases without bound as s increases from 0.

15. Comment on the level sets in the positive orthant in n-dimensional space of

$$\Lambda(x) = \sum_{i=1}^{n} \lambda_i \left[\frac{x_i}{\hat{x}_i} - \ln\left(\frac{x_i}{\hat{x}_i}\right) - 1 \right]$$

Where $\lambda_1, \lambda_2, \ldots, \lambda_n$ are positive constants and $\hat{x} = (\hat{x}_1, \hat{x}_2, \ldots, \hat{x}_n)$ is

in the positive orthant. Use the ray $(\hat{x}_1 + sa_1, \hat{x}_2 + sa_2, \ldots, \hat{x}_n + sa_n)$ where s is
a nonnegative variable and $a_1, a_2, \ldots, a_n$ are constants, not all zero.

16. Use the results of problems 8 through 11 to show that $\hat{x}(t) = (1,1,1)$ is a constant
attractor trajectory for the model (from problem 12)

$$\dot{x}_1 = x_1(-x_1-x_2+2)$$
$$\dot{x}_2 = x_2(2x_1-x_2-x_3)$$
$$\dot{x}_3 = x_3(x_2-x_3)$$

Show that the basin of attraction of $\hat{x}(t)$ is the positive orthant.

17. Show that the constant trajectory $\hat{x}(t) = (100,10,1)$ is an attractor trajectory for
for model of simplistic autotroph-herbivore-carnivore dynamics given by

$$\dot{x}_1 = x_1(-x_1-x_2+110)$$
$$\dot{x}_2 = x_2(.1x_1-.9x_2-x_3)$$
$$\dot{x}_3 = x_3(.1x_2-x_3)$$

Use the function

$$\Lambda(x,t) = \sum_{i=1}^{3} \lambda_i \left[\frac{x_i}{\hat{x}_i} - \ln\left(\frac{x_i}{\hat{x}_i}\right) - 1 \right]$$

with $\lambda_1 = \lambda_2 = \lambda_3 = 1$.

60

ANSWERS TO PROBLEMS

1. $\dot{x}_1 = -.1x_1$

2. $\dot{x}_1 = -x_1+x_2; \quad \dot{x}_2 = -x_1-x_2$

3. $\dot{x}_1 = 2x_1\cos(x_1); \dot{x}_2 = 2;$ and $\dot{x}_3 = x_1x_2x_3$.

4. A Lotus 1-2-3 program is listed below. Note recalculation is rowwise and macro range is A1--A4.

```
A1:  '{goto}B8~+B6+B8~
A2:  '{calc}
A3:  '/xi(B7-B8)>0~/xqA2~
A4:  '/xq
A6:  'delta t
B6:  0.001
A7:  'max t
B7:  1
A8:  'current  t
B8:  +B6+B8
A10: +A22
B10: +B22
A11: +A23
B11: +B23
B12: (F9) @SQRT(A10^2+A11^2)
A13: (+A11)*$B$6
A14: (-A10)*$B$6
A16: +A10+A13
A17: +A11+A14
A19: (+A17)*$B$6
A20: (-A16)*$B$6
A22: +A10+0.5*(A13+A19)
A23: +A11+0.5*(A14+A20)
```

The distances to (0,0) from the trajectory values at t = 1 for the four simulations are:

Δt = .01, forward-difference procedure   1.421373045, error = 7.2×10^{-3}

Δt = .001, forward-difference procedure   1.414921553, error = 7.1×10^{-4}

Δt = .01, double-approximation procedure   1.414213741, error = 1.8×10^{-7}

Δt = .001, double-approximation procedure   1.414213563, error = 10^{-9}

5. Any constant trajectory $\hat{x}(t)$ has $0 = -x_1 - x_2$ and $0 = +x_1 - x_2$. Thus $x_1 = -x_2 =$

$x_2 = 0$. Along an arbitrary trajectory x(t) we have $\dot{\Lambda} = x_1\dot{x}_1 + x_2\dot{x}_2 = x_1(-x_1-x_2) + x_2(x_1-x_2) = -x_1(t)^2 - x_2(t)^2$. The level sets of Λ in (x_1, x_2,t)-space are right circular cylinders with common axis (0,0,t). Thus outside the level cylinder of radius ε about $\hat{x}(t) = (0,0)$, $\dot{\Lambda}$ is less than $-\varepsilon^2$; hence $\beta = \varepsilon^2$.

6. First of all

$$x_1^2(t) + x_2^2(t) = x_1^2(0)\cos^2(t) - 2x_1(0)x_2(0)\sin(t)\cos(t) + x_2^2(0)\sin^2(t)$$

$$+ x_1^2(0)\sin^2(t) + 2x_1(0)x_2(0)\sin(t)\cos(t) + x_2^2(0)\cos^2(t)$$

$$= x_1^2(0)(\cos^2(t) + \sin^2(t)) + x_2^2(0)(\sin^2(t) + \cos^2)(t))$$

$$= x_1^2(0) + x_2^2(0)$$

$$= 1$$

Hence the points $(x_1(t), x_2(t))$ lie on the unit circle for all time. Now

$\dot{x}_1 = x_1(0)(-\sin(t)) - x_2(0)\cos(t) = -x_2(t)$. Similarly, $\dot{x}_2(t) = +x_1(t)$. Since $x_1(t)^2 + x_2(t)^2 = 1$, we also have

$$\dot{x}_1 = (1-x_1^2-x_2^2)x_1 - x_2$$
$$\dot{x}_1 = (1-x_1^2-x_2^2)x_2 + x_1$$

Hence $(x_1(t), x_2(t))$ is actually a trajectory of the dynamical system.

7. For $\lambda > 0$, $(x_1^2+x_2^2-1)^2 = \lambda$ if and only if $x_1^2+x_2^2 = 1 \pm \sqrt{\lambda}$. The negative square root of λ can be used only if $\lambda \leq 1$. Thus the level set associated with $\lambda = 0$ is the unit circle (radius = 1, center = (0,0)); the level set associated with $\Lambda =$ .25 is the pair of circles centered at (0,0) with radii $\sqrt{3/2}$ and $\sqrt{1/2}$; the level set associated with $\lambda = 1$ is the circle centered at (0,0) of radius $\sqrt{2}$ and the point (0,0); the level sets associated with $\lambda = 4$ and $\lambda = 10^6$ are circles centered at (0,0) of radii $\sqrt{3}$ and $\sqrt{1001}$. A suggestive way to think of level sets of $(x_1^2 + x_2^2-1)^2$ is to consider the surface in three-dimensional space consisting of points $(x_1,x_2,(x_1^2+x_2^2-1)^2)$. A sketch of this surface is shown in Fig. 3.7. The surface

is like an infinite bowl with a bump at the bottom. If one imagines partly filling the bowl with water, one can thing of a level set of $(x_1^2 + x_2^2 - 1)^2$ as the edge of the water surface.

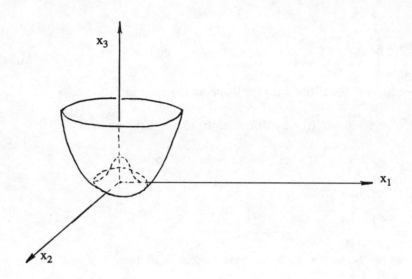

Figure 3.7. The surface of points $(x_1, x_2 \ (x_1^2 + x_2^2 - 1)^2)$.

8. We have

$$\dot{\Lambda} = 2(x_1^2 + x_2^2 - 1)(2x_1\dot{x}_1 + 2x_2\dot{x}_2)$$

$$= 4(x_1^2 + x_2^2 - 1)[(1 - x_1^2 - x_2^2)\,x_1^2 - x_1 x_2 + (1 - x_1^2 - x_2^2)x_2^2 + x_1 x_2]$$

$$= 4(x_1^2 + x_2^2 - 1)(1 - x_1^2 - x_2^2)(x_1^2 + x_2^2)$$

$$= -4(x_1^2 + x_2^2 - 1)^2(x_1^2 + x_2^2)$$

Hence $\dot{\Lambda} = 0$ if and only if $x_1^2 + x_2^2 = 1$ or 0. That is, $\dot{\Lambda} = 0$ on the unit circle and at the origin of two-dimensional space. Elsewhere $\dot{\Lambda} < 0$.

Considering Fig. 3.7 we can think of a typical trajectory of the system as the path of a particle which rolls generally down the sides of the bowl and around the bowl counterclockwise as seen from above. If the particle starts at $(0,0,1)$, the peak of the bump, then it stays at $(0,0,1)$ because $\hat{x}(t) = (0,0)$ is a constant trajectory for the system.

9. Using $\Lambda = x_1^2 + kx_2^2$ we are led to

$$\dot{\Lambda} = 2x_1(-x_1+3x_2) + 2kx_2(-x_1-x_2)$$
$$= -2x_1^2 + 6x_1x_2 - 2kx_1x_2 - 2kx_2^2$$

The best choice for k is evidently $k = 3$, for then $\dot{\Lambda} = -2x_1^2 - 6x_2^2$ (the cross terms x_1x_2 cancel). Thus $\dot{\Lambda} < 0$ except along the constant trajectory $\hat{x}(t) = (0,0)$. Typical level sets of $\Lambda(x_1,x_2,t) = x_1^2 + 3x_2^2$ are then right elliptical cylinders with a common axis, the line $(0,0,t)$. Clearly there exist constant radius cylinders which lie inside or outside any such level set of Λ. Likewise there exist level sets of Λ which lie inside or outside any level cylinder about $\hat{x}(t)$.

Outside a level cylinder of radius ε, $\dot{\Lambda} = -2x_1^2 - 6x_2^2 \leq -2x_1^2 - 2x_2^2 \leq -2\varepsilon^2$.

Hence $\beta = 2\varepsilon^2$ may be used to fulfill condition 3 of the Lyapunov Theorem. Since there is no limit on the size of level sets or level cylinders in these considerations, the basin of attraction must be all of state space-time.

10. Using $\Lambda(x_1,x_2,t) = x_1^2 + 3x_2^2$ again, we are led to

$$\dot{\Lambda} = 2x_1\dot{x}_1 + 6x_2\dot{x}_2$$
$$= 2x_1(-x_1+3rx_2) + 6x_2(-rx_1-x_2)$$
$$= -2x_1^2 + 6x_1x_2r - 6x_1x_2r - 6x_2^2$$
$$= -2x_1^2 - 6x_2^2$$

All the arguments in the solution of the previous problem remain valid, so $\hat{x}(t) = (0,0)$ is again an attractor trajectory. This problem illustrates qualitative type of stability in that $r(x,t)$ is an arbitrary analytic function.

11. Since $\dot{\Lambda} \leq -2x_1^2 - 6x_2^2$, the techniques used in solving the previous two problems apply directly.

12. Straightforward substitution shows $\hat{x}(t) = (1,1,1)$ is a constant trajectory. Using the suggested Λ we have

$$\dot{\Lambda} = 2(1-x_1^{-1})\dot{x}_1 + (1-x_2^{-1})\dot{x}_2 + (1-x_3^{-1})\dot{x}_3$$

$$= 2\left[\frac{x_1-1}{x_1}\right]\dot{x}_1 + \left[\frac{x_2-1}{x_2}\right]\dot{x}_2 + \left[\frac{x_3-1}{x_3}\right]\dot{x}_3$$

$$= 2(x_1-1)\,(-x_1-x_2+2) + (x_2-1)\,(2x_1-x_2-x_3) + (x_3-1)(x_2-x_3)$$

Using the algebraic identities

$$-x_1-x_2+2 = -(x_1-1) - (x_2-1)$$
$$2x_1-x_2-x_3 = +2(x_1-1) - (x_2-1) - (x_3-1)$$
$$x_2-x_3 = (x_2-1) - (x_3-1)$$

we can simplify $\dot{\Lambda}$ to
$$\dot{\Lambda} = -2(x_1-1)^2 - (x_2-1)^2 - (x_3-1)^2$$

Thus Λ decreases along all trajectories except $\hat{x}(t)$.

13. If $x > 0$ and $y = x - \ln(x) - 1$, then $dy/dx = 1 - x^{-1}$ and $d^2y/dx^2 = x^{-2}$. Thus $dy/dx = 0$ if and only if $x = 1$. For all x ($x>0$) $d^2y/dx^2 > 0$, so the function is

concave upward with a minimum at $x = 1$. Since $\lim_{x \to 0^+} y = +\infty$, the function

has a vertical asymptote at $x = 0$. The function is greater than or less than the linear function $x - 1$ according to whether x is less than or greater than 1. A sketch is provided in Fig. 3.8.

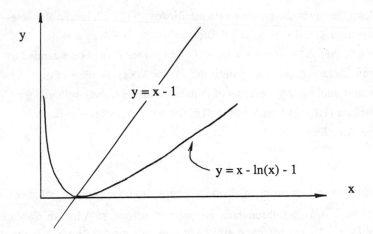

Figure 3.8. A sketch of two functions in (x,y)-space.

14. We have

$$d\Lambda/ds = \sum_{i=1}^{n} \frac{\partial}{\partial x_i} (x_i - \ln(x_i) - 1) \frac{\partial x_i}{\partial s}$$

$$= \sum_{i=1}^{n} (1 - x_i^{-1})(a_i) = \sum_{i=1}^{n} (1 - \frac{1}{1 + sa_i})(a_i)$$

$$= \sum_{i=1}^{n} \frac{sa_i^2}{1 + sa_i}$$

Thus $\Lambda = 0$ and $d\Lambda/ds = 0$ at the point $(1,1,\ldots 1)$ (where $s = 0$). For small values of s (sufficiently small so each $1 + sa_i > 0$) the sign of $d\Lambda/ds$ is positive. Also

$$d^2\Lambda/ds^2 = \sum_{i=1}^{n} \frac{(a_i^2)(1 + sa_i) - (sa_i^2)a_i}{(1 + sa_i)^2}$$

$$= \sum_{i=1}^{n} \frac{a_i^2}{(1 + sa_i)^2}$$

Thus along the ray in the positive orthant, $d^2\Lambda/ds^2 > 0$, so Λ has an absolute minimum at $(1,1,\ldots,1)$. If all $a_i \geq 0$, then clearly $\Lambda \to +\infty$, as $s \to +\infty$. If some $a_i < 0$, then $\Lambda \to +\infty$ as $1 + sa_i \to 0$. Of course, Λ may be extended to a function of state space-time by defining $\Lambda(x) = \Lambda(x,t)$. In view of our observations about Λ, any level set of Λ must lie inside some level cylinder about the line $(1,1,\ldots,1,t)$ and must lie outside some other level cylinder about the same line.

15. As in the previous problem, $d\Lambda/ds > 0$ throughout the positive orthant except at $\hat{x}$. Likewise $d^2\Lambda/ds^2 > 0$ throughout the positive orthant, so Λ has an absolute minimum at $\hat{x}$. Also as before, Λ approaches $+\infty$ as s grows. We may again extend Λ to a function of state space-time, namely $\Lambda(x) = \Lambda(x,t)$. As such, any level set of Λ inside some level cylinder about the line $(\hat{x},t)$ and must lie outside some other level cylinder about the same line.

16. Any point in the positive orthant lies in a level set of Λ as defined in problem 8. Considering the observations in the answers to problems 8 through 11, the Lyapunov Theorem can be used to prove $\hat{x}(t) = (1,1,1)$ is a constant attractor trajectory for the system with the positive orthant as basin of attraction.

17. The suggested Λ is of the form dealt with in problem 11. Along an arbitrary trajectory

$$\dot{\Lambda} = \sum_{i=1}^{3} \frac{\partial\Lambda}{\partial x_i} \dot{x}_i$$

$$= \left(\frac{1}{100} - \frac{1}{x_1}\right)\dot{x}_1 + \left(\frac{1}{10} - \frac{1}{x_2}\right)\dot{x}_2 + \left(1 - \frac{1}{x_3}\right)\dot{x}_3$$

$$= (x_1-100)(-x_1-x_2+110)/100 + (x_2-10)(.1x_1-.9x_2-x_3)/10 + (x_3-1)(.1x_2-x_3)$$

Using the algebraic identities

$-x_1 - x_2 + 110 = -(x_1-100) - (x_2-10)$
$.1x_1 - .9x_2 - x_3 = .1(x_1-100) - .9(x_2-10) - (x_3-1)$
$.1x_2 - x_3 = .1(x_2-10) - (x_3-1)$

we have

$\dot{\Lambda} = -(x_1-100)^2/100 - .9(x_2-10)^2/10 - (x_3-1)^2$

Thus outside the level cylinder about $\hat{x}(t) = (100,10,1)$ of radius ε, $\dot{\Lambda} < -\varepsilon^2/100$.

Considering this inequality, the Lyapunov Theorem, and the answer to problem 11, $\hat{x}(t)$ must be an attractor trajectory for the system.

CHAPTER FOUR
INTRODUCTION TO ECOSYSTEM MODELS

4.1 Brewing Beer and Yeast Population Dynamics

In this chapter we reinforce some of the ideas in Chapter Three, beginning with a simplified version of the biochemistry of brewing.

Beer can be brewed by mixing 50 liters of 25°C water, four kg table sugar, about two liters hop-flavored malt, and some beer yeast or the dregs of a previous batch in a covered stone crock. Within a few days fermentation by a growing yeast population (*Saccharomyces* spp. fungi) results in the release of considerable carbon dioxide. After about two weeks fermentation ceases upon the depletion of sugar and the production of ethanol. Brewing beer in some jurisdictions is illegal.

Initially the yeast population grows and we look at that growth first. The usual model is

$$\dot{x} = rx - rx^2/K = rx(1 - x/K)$$

Here x is the yeast population. The "rx" term is the Malthusian growth term, the multiplication of the yeast population in the abundance of nutrient. The term "$-rx^2/K$" is associated with auto-regulation. The constant K is called the carrying capacity because if x = K then x = 0, that is, if the initial population is K, then the population remains constant. Thus x(t) = K is a (constant) trajectory for the system.

This system is so simple that an explicit solution is possible, namely

$$x(t) = K\left[1 + \left(\frac{K}{x(0)} - 1\right)e^{-rt}\right]^{-1} \qquad (0 < x(0) < K,\ x(0) = K,\ \text{or}\ x(0) > K)$$

The reader should be able to verify using calculus rules that such x(t) are indeed trajectories for the one-dimensional system. Three typical trajectories obtained by computer simulations with the double-approximation procedure are shown in Fig. 4.1. The "S" shaped trajectory with 0 < x(0) < K is called a *logistic growth curve*.

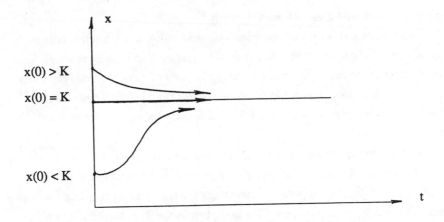

Figure 4.1. Three types of yeast population trajectories.

Using the explicit form of $x(t)$ we can see each trajectory approaches the constant trajectory $x(t) = K$ with time.

Although this model is widely quoted it is easily made more realistic. As all the sugar is consumed, the yeast population, meaning the mass of yeast engaged in fermentation, falls to almost zero. A mass of inactive yeast spores and dead yeast (starved by the depletion of sugar and poisoned by the alcohol) accumulates at the bottom of the crock. This observation suggests that the miniature "ecosystem" in the crock might at any time be characterized by two numbers: x_1 = mass of sugar; and x_2 = mass of yeast engaged in fermentation.

Assuming a "collisions" type mechanism for the rate of sugar depletion leads to

$$\dot{x}_1 = -Ax_1x_2$$

where A is a positive reaction rate constant.

Blending our previous one-dimensional model and the "collisions" mechanism leads to

$$\dot{x}_2 = +Bx_1x_2 - Cx_2^2$$

where B and C are positive reaction rate and auto-regulation constants. The three constants A, B, and C are all in units energy^{-1} time^{-1}.

Of course, if sugar (x_1) and yeast (x_2) are measured in the same units, such as in number of carbon atoms, then A must be greater than B. Some carbon atoms in the reaction go out of the system by becoming part of carbon dioxide or ethanol molecules. It is quite possible to add compartments to our model and conserve carbon, but for our present purpose it is sufficient to deal only with sugar and yeast.

Note that our system functions $-Ax_1x_2$ and $+Bx_1x_2-Cx_2^2$ are continuous throughout the nonnegative orthant. Hence given an initial time t_0 and an initial state ($x_1(t_0)$, $x_2(t_0)$) in the positive orthant, the system evolves according to values a unique trajectory for the system. This is implied by the Trajectory Existence Theorem of the previous chapter. Furthermore, no trajectory can reach the finite boundary of the positive orthant in a finite amount of time. This is implied by the Trajectory Trapping Theorem, using:

$$g_1(x_2,x_2,t) = -Ax_2$$
$$h_1(x_1,x_2,t) = 0$$
$$g_2(x_1,x_2,t) = Bx_1-Cx_2$$
$$h_2(x_1,x_2,t) = 0.$$

So we know a little about typical trajectories of the sugar-yeast system. Solving the system equations explicitly with combinations of "elementary functions" (e^x, polynomials, trigonometric functions) appears to be difficult. However, we can qualitatively characterize all trajectories anyway, using another idea adapted from those of the nineteenth century mathematical wizard M. A. Lyapunov.

Consider the function Λ which gives us a number for any point (x_1,x_2) in state space defined by

$$\Lambda(x_1,x_2) = x_2+Bx_1/A$$

Thus Λ is in the same units as the state variables: energy, fixed carbon, or the equivalent. What is Λ, really? It is just a mathematically useful measure of the current state of the system, as we proceed to demonstrate.

Now for any positive number λ, the *level set of Λ associated with λ* is just the set of all (x_1,x_2) in the nonnegative orthant with $\Lambda(x_1,x_2) = \lambda$. Thus ($x_1,x_2$) is in this level set if $x_2 = +\lambda-Bx_1/A$. (You should be able to verify this algebraically.) Of course, $x_2 = +\lambda-Bx_1/A$ is just an equation for a straight line (an equation of the type $x_2 = mx_1+b$) with negative slope. Different lines (level sets) associated with different λ values are parallel. Typical level sets are shown in Fig. 4.2.

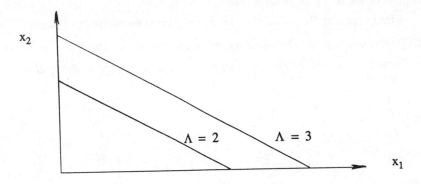

Figure 4.2. Typical level sets of $\Lambda = x_2+Bx_1/A$.

Suppose $(x_1(t),x_2(t))$ is a trajectory for the system. All we know about this trajectory is that along it

$$\dot{x}_1 = -Ax_1x_2$$
$$\dot{x}_2 = +Bx_1x_2-Cx_2^2$$

Along the same trajectory the Lyapunov function has the values

$$\Lambda(x_1(t),x_2(t)) = x_2(t)+Bx_1(t)/A$$

We do not know explicitly what such values are, but we know from the *Chain Rule* of calculus that the rate of change of Λ along the trajectory is

$$\dot{\Lambda} = \dot{x}_2 + B\dot{x}_1/A = Bx_1x_2-Cx_2^2 + B(-Ax_1x_2)/A = -Cx_2^2$$

Perhaps the most popular part of calculus is cancelling. As in this example, there is almost always some cancelling in differentiating Lyapunov functions.

We have defined a function Λ, claimed that it is a "Lyapunov function," and calculated its derivative along a trajectory. Notice that the derivative of Λ is negative throughout the positive orthant, in other words, Λ must be <u>decreasing</u> <u>along</u> <u>all</u> <u>trajectories</u> <u>in</u> <u>the</u> <u>positive</u> <u>orthant</u>. Said another way, our arbitrary trajectory must be puncturing the level sets of Λ "inwardly" relative to the origin.

Each level set of Λ (where $\Lambda = \lambda$) together with the boundary of the positive orthant encloses a triangular patch of the positive orthant. The smaller the value of

λ, the smaller the patch. Since Λ is decreasing along any trajectory we choose, any trajectory we choose must be moving into smaller and smaller patches or must reach the finite boundary of the positive orthant.

On the part of the boundary of the positive orthant consisting of points $(0, x_2)$, $\dot{\Lambda}$ is still negative; hence no trajectory can stop there. However, any point $(x_1, 0)$ on the other part of the boundary is a constant trajectory for the system.

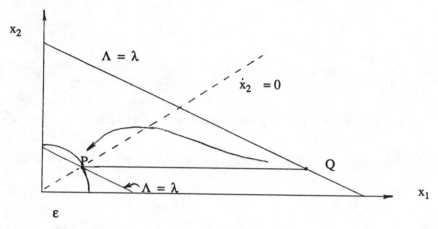

Figure 4.3. A sketch of sugar and yeast system dynamics.

The locus of points in the positive orthant where $\dot{x}_2 = 0$ is given by the relation $x_2 = C^{-1}Bx_1$, the dashed line in Fig. 4.3 with positive slope passing through $(0,0)$. Let $x(0)$ be any point in the positive orthant. Let λ be any number greater than $A^{-1}Bx_1(0) + x_2(0)$ and let ε be any number less than $x_2(0)[1+C^2/B^2]$. The level set $\Lambda = \lambda$ and the set of points satisfying $d(x,0) \le \varepsilon$ (a quarter circle) are shown relative to $x(0)$ in Fig. 4.3. Now the trajectory $x(t)$ starting at $x(0)$ cannot cross downwardly the line PQ because any trajectory on PQ has $\dot{x}_2$ positive.

Similarly $x(t)$ cannot cross the level set $\Lambda = \lambda$ because Λ decreases along all trajectories in the positive orthant. Also, $x(t)$ cannot reach the x_2-axis in finite time because of the Trajectory Trapping Theorem. Outside the quarter circle and above PQ, Λ is not just decreasing along $x(t)$, Λ is decreasing at a rate which cannot approach zero. It follows that $x(t)$ must within a finite time enter the quarter circle area. Moreover, $x(t)$ cannot subsequently escape the triangular patch bounded by the level set $\Lambda = \dfrac{A+C}{A(1+C^2/B^2)^{1/2}} \varepsilon = \lambda'$ and the boundary of the positive orthant.

Since ε and λ' can be arbitrarily small, this proves that x(t) asymptotically approaches the constant trajectory (0,0).

The above analysis based on the Lyapunov function enables us to conclude that any trajectory of the sugar and yeast system starting in the positive orthant is qualitatively like the one shown in Fig. 4.3. Thus any trajectory which starts in the positive orthant must asymptotically approach the origin. Note that the initial behavior of a trajectory starting with much sugar and little yeast is essentially the same as logistic growth in the previous one-dimensional model $\dot{x} = rx(1-x/K)$.

All this illustrates the qualitative Lyapunov approach in differential equation dynamical system modeling. Repeated difference equation calculations could never establish that all trajectories of all systems (A,B,C positive) which start in the positive orthant asymptotically approach (0,0). Moreover, the sugar and yeast model is complicated enough so that the prospect of solving directly for trajectory functions is not attractive. In other words, qualitative geometric analysis is the proper and easy way to determine the qualitative behavior of the model. Of course, quantitative information about a particular trajectory of a particular version of the model could be had by computer simulation.

4.2   Attractor Trajectories

As we have seen, some difference and differential equation dynamical systems have special trajectories to which other trajectories are attracted. General trajectories of the yeast model were attracted to the constant trajectory $x(t) = K$ and general trajectories of the sugar and yeast model were attracted to the constant trajectory $(x_1(t),x_2(t)) = (0,0)$.

We need to formalize the notion of asymptotic approach. Suppose we have two trajectories $x(t)$ and $\hat{x}(t)$ of a dynamical system with initial values $x(t_0)$ and $\hat{x}(t_0)$. We say that $x(t)$ asymptotically approaches $\hat{x}(t)$ if for any positive number ε there exists a finite time interval T such that $d(x(t), \hat{x}(t)) < \varepsilon$ for all $t > T$; T is determined by $x(t_0),x(t_0)$, and ε, not by t_0 itself.

The concept of attractor trajectory may be defined as follows. Recall that in n-dimensional space the distance $d(x,y)$ between two points x and y is

$$d(x,y) = \left[\sum_{i=1}^{n} (x_i-y_i)^2 \right]^{1/2}$$

This is just the n-dimensional extension of the Pythagorean Theorem (see problems 1.5 through 1.10). The reader should have no difficulty showing that the distance between $(1,2,3,0,0,1,-1)$ and $(1,2,2,1,0,-1,2)$ is $\sqrt{15}$. The *triangle inequality* of linear algebra is for any x,y, and z in n-dimensional space

$$d(x,y)+d(y,z) \geq d(x,z)$$

We shall call a certain trajectory $x(t)$ of a dynamical system an *attractor trajectory* with respect to a certain basin of attraction S in state space if the following two properties hold:

1. any trajectory $x(t)$ which lies in S at some time asymptotically approaches $\hat{x}(t)$; and

2. for a number $\varepsilon > 0$ there exists another number $\delta > 0$ such that $d(x(t_0), \hat{x}(t_0)) < \delta$ implies $d(x(t),\hat{x}(t)) < \varepsilon$ for all $t > t_0$, with δ determined by ε only.

Roughly speaking, $\hat{x}(t)$ is an attractor trajectory relative to a *basin of attraction* S if any other trajectory passing through S ultimately approaches $\hat{x}(t)$ and if trajectories can be guaranteed to stay close to $\hat{x}(t)$ just by requiring that they start close to $\hat{x}(t)$.

Let us illuminate the concept of attractor trajectory with an example. In section 2.1 we considered

$$x(t+\Delta t) = x(t)+[-.1x^2(t)-2x(t)y(t)+2.1x(t)]\Delta t$$
$$y(t+\Delta t) = y(t)+[2x(t)y(t)-.1y^2(t)-1.9y(t)]\Delta t$$

as a first step toward modeling a predation interaction. Consider the analogous differential equation dynamical system

$$\dot{x}_1 = -.1x_1^2-2x_1x_2+2.1x_1$$
$$\dot{x}_2 = 2x_1x_2-.1x_2^2-1.9x_2$$

The subscript notation (x_1 = prey, x_2 = predator) is used here because we want to be able to generalize readily to systems of higher dimension. Simulations in the previous chapter suggest that the trajectory of this differential equation dynamical system starting at $(2,1)$ should asymptotically approach the constant trajectory $(1,1)$.

It turns out that the Lyapunov function we need this time is

$$\Lambda(x_1,x_2) = \sum_{i=1}^{2} x_i - \ln(x_i) - 1$$

Calculus and analytic geometry can be used to show that the level sets of Λ in the positive orthant of (x_1,x_2)-space look like the simple closed curves shown in Fig. 4.4. For any point $x(0)$ in the positive orthant, $\Lambda(x(0))$ is less than some finite (positive) number λ, that is, $x(0)$ lies inside the level set $\Lambda = \lambda$. Furthermore, for any level set there exist circles centered at $(1,1)$, one of which lies completely inside the level set, while the other lies completely outside the level set.

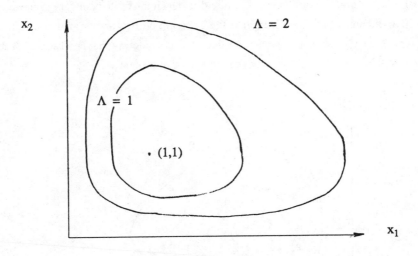

Figure 4.4. **Typical level sets of** $\Lambda = \displaystyle\sum_{i=1}^{2} x_i - \ln(x_i) - 1.$

Admittedly it does take some mathematical effort to verify that the level sets of Λ are as advertised in Fig. 4.4. But developing the curves is far easier than finding approximate trajectory functions.

Using the Chain Rule again, let us calculate the rate of change of Λ along some fixed but arbitrary trajectory $x(t)$ in the positive orthant. We have with some algebraic manipulations

$$\dot{\Lambda} = \dot{x}_1 - \dot{x}_1/x_1 + \dot{x}_2 - \dot{x}_2/x_2$$

$$= (1-1/x_1)\dot{x}_1 + (1-1/x_2)\dot{x}_2$$
$$= (x_1-1)(-.1x_1-2x_2+2.1) + (x_2-1)(2x_1-.1x_2-1.9)$$
$$= (x_1-1)(-.1x_1+.1) + (x_1-1)(-2x_2+2) + (x_2-1)(-.1x_2+.1) + (x_2-1)(2x_1-2)$$

Cancelling the second and fourth summands gives

$$= -.1(x_1-1)^2 -.1(x_2-1)^2 = -.1d(x,(1,1))^2$$

Therefore $\dot{\Lambda}$ is negative everywhere in the positive orthant of (x_1,x_2)-space except at one point, $(1,1)$.

Notice that the Trajectory Trapping Theorem implies that no trajectory starting in the positive orthant can in finite time reach the finite boundary of the positive orthant. (In the application of the Theorem both h_1 and h_2 are zero.) However, we already know even more than that, considering that level sets of Λ are punctured inwardly by all trajectories in the positive orthant.

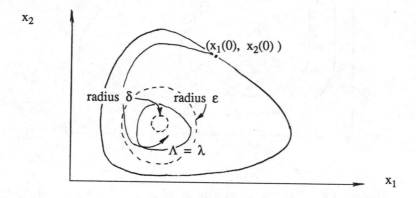

Figure 4.5. A Lyapunov construction which can be used to show (1,1) is an attractor trajectory for the predator-prey system.

Let us show carefully that $(1,1)$ is an attractor trajectory. Let ε be any positive number and let λ be a positive number so that the level set $\Lambda = \lambda$ is entirely inside the circle of radius ε centered at $(1,1)$. Suppose a point $x(0)$ is inside the positive orthant but outside the circle, as shown in Fig. 4.5. Consider the trajectory $x(t)$ which starts at $x(0)$. Let λ_0 be the value of Λ at $x(0)$. Notice that outside the circle Λ must decrease; in fact, $\dot\Lambda$ along $x(t)$ is not just negative, but $\dot\Lambda < -.1\varepsilon$. Therefore, since Λ at $x(0)$ is λ_0, since Λ on the circle is greater than λ, and since $\dot\Lambda < -.1\varepsilon$, at most $(\Lambda_0 - \lambda)/(.1\varepsilon)$ time units can elapse before $x(t)$ enters the circle. Recall that ε can be arbitrarily small. Thus any trajectory of the system enters an artibrarily small circular region about $(1,1)$ after a finite time interval. The length of the time interval depends on the starting point of the trajectory and the diameter of the region.

Keeping ε and λ as above, choose $\delta > 0$ so that the circle of radius δ centered at $(1,1)$ lies inside the level set $\Lambda = \lambda$. After a slightly longer time interval, $x(t)$ must actually enter the smaller circle and can never thereafter cross the level set $\Lambda = \lambda$, a level set inside the circle of radius ε. Thus $x(t)$ asymptotically approaches $(1,1)$. This establishes the first of the two requirements $(1,1)$ must fulfill to be an attractor trajectory.

We can confirm that $(1,1)$ fulfills the second requirement as follows. Let ε be any positive number. Let λ be a positive number so that the level set $\Lambda = \lambda$ lies inside the circle of radius ε centered at $(1,1)$. Let δ be the radius of a circle centered at $(1,1)$ which lies inside the level set $\Lambda = \lambda$. The circles and level set are shown in Fig. 4.5. Considering the rate of change of Λ along trajectories, it follows that no trajectory $x(t)$ which starts inside the smaller circle can cross the level set $\Lambda = \lambda$. This implies that $x(t)$ can be guaranteed to stay within ε of $(1,1)$ just by requiring that $x(t)$ start within δ of $(1,1)$. Since ε can be arbitrarily small, we have established the second of the two requirements that $(1,1)$ must fulfill to be an attractor trajectory.

In dynamical systems theory the existence of an attractor trajectory is but one of many technical definitions of stability. Attractor trajectories and, as we shall see later, their cousins, attractor regions, are cogent ways of formalizing what ecologists mean by stability.

Ecologists observe patterns of ecosystem development, the orderly, more or less predictable changes over time of ecosystems in response to a large physical disturbance. The succession of plants in a forested area following a fire is a familiar example of an *ecosystem development sere*. A specific case would be the recovery of plant communities in the boreal forest zone of Canada following a fire, as sketched at the beginning of Chapter One. The general succession of plants is easy for someone familiar with the boreal forest to predict; only the details of plant location and growth rate are to some degree random.

Suppose an experimenter were to cut down all saplings and brush from a one hectare plot in a large burned area, say, one year after the burn. It is likely that with the passage of time the experimental plot would be less and less distinguishable.

To put things differently, the recovering forest thought of as energy compartments and so as a dynamical system, would pursue a trajectory which would approach the trajectories of nearby control plots.

Attractor trajectories are not really very special. In the simple predator-prey model above, all trajectories in the positive orthant are attractor trajectories. The trajectory (1,1) is distinguished just by being a constant trajectory as well. Likewise, in studying an ecosystem development sere, ecologists might look not so much for a special attractor trajectory as for the convergence of various trajectories to a common *climax pattern* independent of initial state details. In the boreal forest fire example, initial state details would entail local details of fire severity. Fire severity in a few hectares or even less can range from scorched moss and litter with undamaged trees to almost complete loss through combustion of organic matter excepting tree boles and various root systems.

In mathematics--but not in realistic ecosystem models--an attractor trajectory might have constant values (constant over time). A nonconstant attractor trajectory, on the other hand, can be thought of as a sort of moving target (moving through state space) or, perhaps, a moving black hole, sucking in all nearby trajectories.

4.3   Derivatives of System Functions

In modeling ecosystems we might know of a particular trajectory, for example the observed, nonconstant state vector of an undisturbed ecosystem, which should be an attractor trajectory. It sometimes happens that the model must not only have such a trajectory but must be to some extent built around that trajectory.

We need to recall a little more calculus. Recall that a partial derivative of a function of several variables with respect to one particular variable is just the ordinary derivative of the function obtained by treating all other variables as constants. For example, if $f(x_1,x_2,x_3,t) = x_1 + x_1x_2 + x_3\sin(t)$, then the partial derivatives of f are

$$\frac{\partial f}{\partial x_1} = 1+x_2; \quad \frac{\partial f}{\partial x_2} = x_1; \quad \frac{\partial f}{\partial x_3} = \sin(t); \quad \frac{\partial f}{\partial t} = x_3\cos(t)$$

Like ordinary differentiation, partial differentiation creates a new function from an old function. Thus the value of $\partial f/\partial x_1 = 1+x_2$ at the point $(x_1,x_2,x_3,t) = (1,2,3,4)$ is 3. The reader should be able to verify that $\partial f/\partial x_3$ at $(1,0,0,\pi/2)$ is 1. A second partial derivative is just a partial derivative of a partial derivative. For our example function f, $\partial^2 f/\partial x_1\partial x_2$ is the constant function 1. The function $\partial^2 f/\partial x_3\partial t$ is just cos(t).

Let us denote a given, observed, or expected trajectory by $\hat{x}(t)$, anticipating that it will turn out to be an attractor trajectory.

An important theorem in dynamical systems theory concerns partial derivatives of system functions evaluated on special trajectories like $\hat{x}(t)$. That is, given

$$\dot{x} = f(x,t)$$

we can use mathematics to say something about the stability of the system around $\hat{x}(t)$. We first compute the n^2 functions of time (possibly constants)

$$\frac{\partial f_i}{\partial x_j}(\hat{x}(t),t)$$

To illustrate the computation of such functions, let us return to the simple predation model

$$\dot{x}_1 = -.1x_1^2 - 2x_1x_2 + 2.1x_1$$
$$\dot{x}_2 = 2x_1x_2 - .1x_2^2 - 1.9x_2$$

The reader should be able to verify that

$$\frac{\partial f_1}{\partial x_1} = -.2x_1 - 2x_2 + 2.1$$

$$\frac{\partial f_1}{\partial x_2} = -2x_1$$

$$\frac{\partial f_2}{\partial x_1} = 2x_2$$

$$\frac{\partial f_2}{\partial x_2} = 2x_1 - .2x_2 - 1.9$$

If we evaluate these four functions on the trajectory $\hat{x}(t) = (1,1)$ we get the four constants -.1, -2, 2, -.1. We can display these numbers in the *matrix*

$$\begin{pmatrix} -.1 & -2 \\ 2 & -.1 \end{pmatrix}$$

Much can be deduced from these four numbers, or, more precisely, from numbers computed from the matrix. This will give us a new tool in thinking about stability.

4.4 The Linearization Theorem

A *linear function* f of several variables $x_1, x_2,...,x_n$ is, of course, any sum

$$f(x) = a_0 + a_1 x_1 + a_2 x_2 + ... + a_n x_n$$

where $a_0, a_1,...,a_n$ are numbers. If the *Taylor series* of a function of several variables exists and converges for all x, then the function is called *analytic*. If a function is the sum of a finite number of products of (nonnegative integer) powers of variables, the function is called a *multinomial*. Any multinomial is analytic over all of state space. Any fraction of multinomials is analytic over that portion of state space in which the denominator is nonzero. For example,

$$f(x,t) = 5 + x_1 x_2 + x_2 x_3^2 + x_3 + t$$

is a multinomial in $x = (x_1, x_2, x_3)$ and t. Further details of such notions from calculus may be found in the Appendix of Chapter Three.

A linear dynamical system is a dynamical system with system functions which are linear in x. For example, the following is a simple *linear dynamical system*

$$\dot{x}_1 = -.1x_1 - 2x_2$$
$$\dot{x}_2 = 2x_1 - .1x_2$$

In fact, this system is the linearization of the predator-prey dynamical system

$$\dot{x}_1 = -.1x_1^2 - 2x_1 x_2 + 2.1x_1$$
$$\dot{x}_2 = 2x_1 x_2 - .1x_2^2 - 1.9x_2$$

about its trajectory $\hat{x}(t) = (1,1)$. Generally the *linearization of* $\dot{x} = f(x,t)$ *about a trajectory* $\hat{x}(t)$ is defined as the linear dynamical system

$$\dot{x}_i = \sum_{j=1}^{n} \frac{\partial f_i}{\partial x_j} (\hat{x}(t),t) x_j$$

with *linear approximation matrix* $A_{ij}(t) = \frac{\partial f_i}{\partial x_j} (\hat{x}(t),t)$. If each partial derivative on

$\hat{x}(t)$, that is, each $\dfrac{\partial f_i}{\partial x_j} (\hat{x}(t),t)$ is actually a constant (independent of time) as in

our example, the linearization is called *autonomous*.

In plain terms, the linear approximation matrix reveals the immediate response of the system to a perturbation. If the system is operating on an attractor trajectory and if $A_{ij} > 0$, then an immediate effect of increasing x_j is an increase in x_i. Likewise, if the system is operating on an attractor trajectory and if $A_{ij} < 0$, then an immediate effect of increasing x_j is a decrease in x_i.

Notice that $0 = (0,0,...,0)$ is always a constant trajectory for a linear dynamical system which arises as a linearization. In fact, 0 in the linear system is analogous to $\hat{x}(t)$ in the general system.

The Linearization Theorem stated below is included in this book because for some systems it enables us to tell whether or not a given trajectory $\hat{x}(t)$ is an attractor trajectory. As with the Lyapunov theory, we need not try to find explicitly typical trajectories near $\hat{x}(t)$--a task that is generally impossible anyway.

Linearization Theorem. Suppose system functions f(x,t) are analytic in x and t. Suppose also:

1. 0 is an attractor trajectory for the linearization of $\dot{x} = f(x,t)$ about $\hat{x}(t)$; and

2. there exists a number N such that $\left| \dfrac{\partial f_i}{\partial x_j} (\hat{x}(t),t) \right| \leq N$ for

all i,j,t. Then $\hat{x}(t)$ is an attractor trajectory for the full system. Some open neighborhood of $\hat{x}(t)$ in state space-time is a basin of attraction for $\hat{x}(t)$.

A proof and details of implications of this theorem can be found in [W, Chapter 5].

To illustrate the Linearization Theorem, let's go back to the one-dimensional yeast model

$$\dot{x} = rx(1-x/K)$$

Of course, we already know the constant trajectory $\hat{x}(t) = K$ is an attractor trajectory, but let us see how the Linearization Theorem can be applied.

First of all, the system function $f(x) = rx(1-x/K)$ is certainly a multinomial and is therefore analytic. Also $\partial f/\partial x$ at the constant trajectory $\hat{x}(t) = K$ is just $r-2r\hat{x}(t)/K = r-2r = -r$. Thus the linearization of the dynamical system $x = rx(1-x/K)$ is

$$\dot{x} = -rx$$

To apply the Linearization Theorem we must ask:

1. Is 0 an attractor trajectory for $\dot{x} = -rx$?
2. Does there exist a number N such that $|-r| \le N$?

The linear system $\dot{x} = -rx$ has an explicit form for its trajectories, namely, $x(t) = x(0)\exp[-rt]$. Therefore as $t\to\infty$, any $x(t)$ asymptotically approaches 0. We can guarantee that any $x(t)$ stays within ε of 0 simply by requiring that it starts within $\delta = \varepsilon$ of 0. So 0 is an attractor trajectory.

Question 2 is trivial; $|-r| = r$, so $N = r$ suffices.

In view of these observations and the Linearization Theorem, $x(t) = K$ must an attractor trajectory for the full system $\dot{x} = rx(1-x/K)$.

An important limitation of the Linearization Theorem is that nothing is said about the size of the basin of attraction S. By contrast, the existence of a Lyapunov function generally can be used to estimate the size of S. In the case of the yeast model, the Lyapunov function $\Lambda(x) = (x/K)-\ln(x/K)$ can be used to show that S may be taken as the set of all positive x.

Suppose a model of boreal forest energy flow includes compartments for: (1) jack pine needles; (2) spruce grouse; (3) snowshoe hare; and (4) great horned owl. It happens that grouse and hare eat pine needles in winter and great horned owls eat grouse and hare whenever they can be caught, including winter.

The full model would be of the form $\dot{x} = f(x,t)$. Based on field observations we might have an idea of "normal" trajectory values $\hat{x}(t)$, so the linearization of the system about $\hat{x}(t)$ could in principle be computed.

The n^2 functions of time $\dfrac{\partial f_i}{\partial x_j}(\hat{x}(t),t)$ are what ecologists call the entries in the *community matrix* of the system. Let us label the community matrix $A(t)$, an $n \times n$ matrix of functions of time with $\dfrac{\partial f_i}{\partial x_j}(\hat{x}(t),t)$ in *row i, column j*. Thus we abbreviate the linearization as

$$\dot{x}_i = \sum_{j=1}^{n} A_{ij}(t)x_j$$

or simply $\dot{x} = A(t)x$.

Now in our supposed boreal forest model we would expect $A_{12}(t)$ to have negative values, at least for winter daytime values of t. This means that perturbing the system away from $\hat{x}(t)$, say, by somehow increasing the grouse population, would at least temporarily cause the energy level in the pine needle compartment to drop below "normal," below $\hat{x}_1(t)$. The reader should be able likewise to reason that $A_{21}(t) > 0$, $A_{13}(t) < 0$, and $A_{31}(t) > 0$. Likely perturbing the hare population away from $x_3(t)$ would have no direct effect on the grouse population, that is, $A_{23}(t) \cong 0$.

Much of field ecology amounts to trying to estimate system functions. Such problems only occasionally yield to the ingenuity and diligence of field ecologists. But often the signs (+, -, or 0) of the community matrix entries are known with certainty. It sometimes turns out that just knowing the signs of entries in $A(t)$ is sufficient to guarantee that 0 is an attractor trajectory for $\dot{x} = A(t)x$ and so, by the Linearization Theorem, that $\hat{x}(t)$ is an attractor trajectory for the full system $\dot{x} = f(x,t)$ (see Chapter 6). The essence of theoretical systems ecology, using mathematical ideas to try to guess patterns of ecosystem organization, is well illustrated by the linearization approach.

4.5 The Hurwitz Stability Test

In this section we shall study a test for the stability of the linear dynamical system with constant coefficients

$$\dot{x}_i = \sum_{j=1}^{n} A_{ij}\, x_j$$

or, in abbreviation, $\dot{x} = Ax$. This linear dynamical system is obviously a long way from ecosystem reality, but one must start somewhere in learning about multi-dimensional systems. For details of matrix algebra and dynamical systems concepts used in this section, see [W].

To be more specific, we shall study a mathematical procedure called the *Hurwitz test*; it is a series of computations using the n^2 numbers in A and numbers derived from those numbers. In certain cases the test determines whether or not 0 is an attractor trajectory for $\dot{x} = Ax$ [W, pp. 66-71].

How useful is the Hurwitz test? Hand calculation is feasible for models with as many as four or five compartments. Beyond that, computers can be used. But the number of calculations for the general problem with plenty of nonzero entries in A grows exponentially with n, the dimension number. Even with computers the Hurwitz test is too long for higher dimensions, say, $n \geq 20$. In fact, there is presently no known computationally feasible means of finding out whether 0 is an

attractor trajectory for $\dot{x} = Ax$ when A is an n×n matrix with many nonzero entries and n >> 20. So the point of learning about the Hurwitz test is really to gain experience with the behavior of representative community matrix problems up to dimension five.

First we need to recall some linear algebra. An *n×n matrix A* consists of n^2 numbers (for our purposes, always real numbers) arranged in n rows and n columns. A *permutation* of the n integers 1,2,3,...,n is simply a reordering of the integers, for example, n,n-1,...,2,1. Any permutation can be rendered into natural order by a sequence (not unique) of steps, each step being the interchange of two adjacent integers. If the number of steps in such a sequence is even (including zero), the *sign of the permutation* is +1. Otherwise the sign of the permutation is -1. Even though the sequence of steps needed to convert a given permutation into natural order is not uniquely determined, the sign of the permutation is. The reader should be able to verify that the sign of the permutation 2,1,4,6,5,3,7 is -1.

We also recall that the *determinant* of an n×n matrix A is defined as

$$\det A = \sum_{i \in \text{ permutations of } 1,2,\ldots,n} \text{sign}(i)\, A_{1i_1} A_{2i_2} \ldots A_{ni_n}$$

where the sum is carried out over all possible permutations of $1,2,3,\ldots,n$. Thus if A is 1×1, then

$$\det A = A_{11}$$

If A is 2×2, then

$$\det A = A_{11} A_{22} - A_{12} A_{21}$$

If A is 3×3, then

$$\det A = A_{11}A_{22}A_{33} + A_{12}A_{23}A_{31} + A_{13}A_{21}A_{32} - A_{11}A_{23}A_{32} - A_{12}A_{21}A_{33} - A_{13}A_{22}A_{31}$$

The number of summands in det A is *n factorial*, $n! = n\cdot(n-1)\cdot(n-2)\ldots3\cdot2\cdot1$. For larger values of n, say $n = 100$, calculating det A by brute force is not feasible. Of course, there are tricks which can be used for certain types of matrices; type is determined by how many and where zeroes occur. Finding and applying such tricks is part of the business of mathematics.

A *polynomial* in the single variable z is a finite sum of multiples of (nonnegative integer) powers of z. A polynomial is a special case of a multinomial.

The $n\times n$ *identity matrix* I has +1 in each I_{ii} (*main diagonal*) entry and 0 in each I_{ij}, $i\neq j$ (*off-diagonal*) entry. Thus the 3×3 identity matrix is

$$\begin{pmatrix} 1 & 0 & 0 \\ 0 & 1 & 0 \\ 0 & 0 & 1 \end{pmatrix}$$

By zI is meant the matrix with the variable z in each main diagonal entry and 0 in each off-diagonal entry.

The *characteristic polynomial* of an $n\times n$ matrix A is the determinant of the $n\times n$ matrix zI-A where $(zI-A)_{ii} = z - A_{ii}$ and $(zI-A)_{ij} = -A_{ij}$, $i\neq j$. The reader should be able to show that the characteristic polynomial of

$$\begin{pmatrix} 1 & 0 & 2 \\ 2 & 0 & 0 \\ 0 & 0 & 3 \end{pmatrix}$$

is $(z-1)\cdot(z-0)\cdot(z-3)+0\cdot0\cdot0+(-2)\cdot(-2)\cdot0-(z-1)\cdot0\cdot0-0\cdot(-2)\cdot(z-3)-(-2)\cdot(z-0)\cdot(0) = z^3 - 4z^2+3z$.

The *Fundamental Theorem of Algebra* as applied in this section tells us that any polynomial

$$p(z) = z^n+p_{n-1}z^{n-1}+...+p_2z^2+p_1z+p_0$$

where $p_0,p_1,p_2,...$ are real numbers always has n *roots* $r_1,r_2,...,r_n$; each root is a number (possibly complex) satisfying $p(r_i) = 0$. Put another way,

$$p(z) = (z-r_1) (z-r_2) ... (z-r_n).$$

The *eigenvalues* of an n×n matrix A are the roots of the characteristic polynomial of A.

A fundamental result of the theory of linear dynamical systems is that 0 is an attractor trajectory for $\dot{x} = Ax$ (with basin of attraction S being all of n-dimensional space) if and only if each eigenvalue of A has negative real part, that is, either is a negative real number or is a complex number with negative real part [W, p. 49].

If a linear approximation matrix (constant community matrix) has even one eigenvalue which has positive real part, then neither 0 for the linear approximation system nor $\hat{x}(t)$ for the full nonlinear system is an attractor trajectory [W, p. 129]. Thus a purely algebraic test can sometimes determine the qualitative performance of a dynamical system.

The Hurwitz test does not involve calculation of the actual eigenvalues of A. Rather, the Hurwitz test requires the calculation of the determinants of certain 1×1, 2×2, 3×3,..., n×n matrices the entries of which are the numbers 1, p_{n-1}, p_{n-2},..., p_1, and p_0. An explicit list of these *Hurwitz matrices* follows.

$$H_1 = (p_{n-1})$$

$$H_2 = \begin{pmatrix} p_{n-1} & p_{n-3} \\ 1 & p_{n-2} \end{pmatrix}$$

$$H_3 = \begin{pmatrix} p_{n-1} & p_{n-3} & p_{n-5} \\ 1 & p_{n-2} & p_{n-4} \\ 0 & p_{n-1} & p_{n-3} \end{pmatrix}.$$

$$H_4 = \begin{pmatrix} p_{n-1} & p_{n-3} & p_{n-5} & p_{n-7} \\ 1 & p_{n-2} & p_{n-4} & p_{n-6} \\ 0 & p_{n-1} & p_{n-3} & p_{n-5} \\ 0 & 1 & p_{n-2} & p_{n-4} \end{pmatrix}$$

and so on up to H_n. Here p_j is defined to be 0 for negative j.

Hurwitz Stability Test. Let $\dot{x} = Ax$ denote a linear dynamical system with constant trajectory 0. Every Hurwitz matrix has positive determinant if and only if every eigenvalue of A has negative real part and 0 is an attractor trajectory for $\dot{x} = Ax$ [W, p. 66].

It can be shown that each Hurwitz matrix can have positive determinant only if each $p_i > 0$, $i = 0,1,2,...,n-1$. If some such $p_i < 0$, then some eigenvalue of A has positive real part and 0 cannot be an attractor trajectory for $\dot{x} = Ax$; if in addition A is a linear approximation matrix about $\hat{x}(t)$ and some such $p_i < 0$, then $\hat{x}(t)$ cannot be an attractor trajectory.

For low values of n the Hurwitz Test can be rewritten as follows. Each Hurwitz matrix has positive determinant if and only if

n = 1:	$p_0 > 0$
n = 2:	$p_0, p_1 > 0$
n = 3:	$p_0, p_1, p_2 > 0$; $p_2 p_1 - p_0 > 0$
n = 4:	$p_0, p_1, p_2, p_3 > 0$; $p_3 p_2 p_1 - p_1^2 - p_0 p_3^2 > 0$

Let us now go through a problem to illustrate all the above. Suppose in a three-dimensional model species in compartments 1 and 2 are preyed upon by the species in compartment 3. Suppose these interactions are given by the model

$$\dot{x}_1 = 11(1+x_2)^{-1} x_1 - x_1 x_3$$
$$\dot{x}_2 = 11x_2 - x_2 x_3 - x_2^2$$
$$\dot{x}_3 = .1x_1 x_3 + .1x_2 x_3 - 2x_3^2$$

The reader should be able to verify that for a fixed value for x_1, increasing x_2 reduces the energy uptake term $11(1+x_2)^{-1}x_1$ in $\dot{x}_1$. Thus species 2 somehow inhibits energy uptake by species 1. In the positive orthant the system functions are analytic and $x(t) = (10,10,1)$ is a constant trajectory for the system.

The matrix of partial derivatives of the system functions is

$$\begin{pmatrix} 11(1+x_2)^{-1}-x_3 & -11x_1(1+x_2)^{-2} & -x_1 \\ 0 & 11-x_3-2x_2 & -x_2 \\ .1x_3 & .1x_3 & .1x_1+.1x_2-4x_3 \end{pmatrix}$$

At the constant trajectory $x(t) = (10,10,1)$ we have as linear approximation matrix

$$A = \begin{pmatrix} 0 & -10/11 & -10 \\ 0 & -10 & -10 \\ .1 & .1 & -2 \end{pmatrix}$$

The characteristic polynomial is then

$$\det \begin{pmatrix} z & +10/11 & +10 \\ 0 & z+10 & +10 \\ -.1 & -.1 & z+2 \end{pmatrix} = z^3+12z^2+22z+100/11$$

Therefore $p_0 = 100/11$, $p_1 = 22$, $p_2 = 12$ (all positive) and $p_2 p_1 - p_0 = 12 \cdot 22 - \frac{100}{11} > 0$. Using the Hurwitz Stability Test we conclude that $(0,0,0)$ is an attractor trajectory for the linear system $\dot{x} = Ax$. The full system functions are analytic in the positive orthant and the entries in $A(t) = A$ are certainly bounded (in fact, constant). It follows by the Linearization Theorem that the constant trajectory $\hat{x}(t) = (10,10,1)$ is an attractor trajectory for the full system. The basin of attraction of $(10,10,1)$ is not specified, however.

REFERENCE

[W] J. L. Willems, *Stability Theory Dynamical Systems,* Nelson, London, 1970.

PROBLEMS

1. Find the constant trajectories for the two-dimensional system

$$\dot{x}_1 = (x_1^2 - 1)x_2$$
$$\dot{x}_2 = x_1(x_2^2 - 1)$$

2. Consider the two-dimensional system

$$\dot{x}_1 = -x_1 - x_2$$
$$\dot{x}_2 = +x_1 - x_2$$

Describe the set of all points in state space where $\dot{x}_1 = 0$. Describe the set of all points in state space where $\dot{x}_2 = 0$. Show $x(t) = (e^{-t} \cos(t), e^{-t} \sin(t))$ is a trajectory for this system. Sketch $x(t)$.

3. Show $x(t) = (\frac{1}{2}t^2 e^{-t} + t e^{-t} + e^{-t}, t e^{-t} + e^{-t}, e^{-t})$ is a trajectory for the dynamical system

$$\dot{x}_1 = -x_1 + x_2$$
$$\dot{x}_2 = -x_2 + x_3$$
$$\dot{x}_3 = -x_3$$

Find $x(0)$. Using a calculator find $x(1)$, $x(10)$, and $x(100)$.

4. Let $x(t)$ be a trajectory for the one-dimensional system $\dot{x} = rx(1-x/K)$. If $x(0) > K$, show $d(x(t),K)$ decreases so long as $x(t) > K$. If $0 < x(0) < K$, show $d(x(t),K)$ decreases so long as $0 < x(t) < K$.

5. Use calculus to graph qualitatively the function $f(x)$ of one variable given by $f(x) = x/K - \ln(x/K) - 1$, where K is a positive constant.

6. Use $\Lambda = x/K - \ln(x/K) - 1$ to show $\hat{x}(t) = K$ is an attractor trajectory for the dynamical system $\dot{x} - rx(1-x/K)$.

7. Let a function f be defined by $f(x,t) = x_1 + x_1 x_2 x_3 + x_3 \cos(t)$. Calculate

$\partial f/\partial x_2$ at $(1,1,1,0)$, $\partial f/\partial x_3$ at $(0,0,0,0)$, $\partial^2 f/\partial x_1 2x_3$ at $(0,1,0,\pi)$, and $\partial f/\partial t$ at $(0,1,1,\pi/2)$.

8. Let a function f of one variable be defined by $f(x) = 1-x+x^2-x^3$. Find f(2) and a line with the slope df/dx at $x = 2$ passing through the point $(2,f(2))$. Sketch the function and the line in $(x,f(x))$-space.

9. Let a function f of two variables be defined by $f(x_1,x_2) = x_1^2+x_1x_2-2x_2^2$. Show the linear approximation of f near $(1,1)$ given by

$$f(x_1,x_2) \cong f(1,1) + \frac{\partial f}{\partial x_1}(1,1) \cdot (x_1-1) + \frac{\partial f}{\partial x_2}(1,1) \cdot (x_2-1)$$

is $3x_1-3x_2$. Evaluate the linear approximation at $(1.1,1.1)$ and compare the result with $f(1.1,1.1)$.

10. Let $f(x_1,x_2,x_3,t) = x_1+x_1x_2x_3+x_3 \cos(t)$. Show the linear approximation of f near $(0,0,1,\pi/2)$ is $x_1-t+\pi/2$. Evaluate the linear approximation at $(.1,.1,1.1,.1+\pi/2)$ and compare with $f(.1,.1,1.1,.1+\pi/2)$.

11. Let $f(x_1,x_2,x_3) = x_1x_2+\ln(1+x_2^2) -\ln(2)$. Calculate the linear approximation of f at $(0,1,0)$. Evaluate the linear approximation at $(.1,1.1,.1)$ and compare with $f(.1,1.1,.1)$. Next evaluate the approximation at $(.01,1.01,.01)$ and compare with $f(.01,1.01,.01)$.

12. Suppose a model of two competing autotrophs and a herbivore is given by

$$\dot{x}_1 = x_1\left[1+.1(1+x_2)^{-1} -.1x_1-2x_3\right]$$
$$\dot{x}_2 = x_2\left[2+.1(1+2x_1)^{-1} -.2x_2-2x_3\right]$$
$$\dot{x}_3 = x_3(x_1+1.2x_2-x_3)$$

Identify which variable corresponds to the herbivore. What does the Trajectory Trapping Theorem say about this model?

13. Suppose a model of two competing autotrophs and a herbivore is given by

$$\dot{x}_1 = x_1 - .001x_1^2 - .001x_1x_2 - .01x_1x_3$$
$$\dot{x}_2 = x_2 - .0015x_2x_1 - .001x_2^2 - .001x_2x_3$$
$$\dot{x}_3 = -x_3 + .005x_3x_1 + .005x_3x_2$$

Identify which variable corresponds to the herbivore. What does the Trajectory Trapping Theorem say about this model? Michael Gilpin has shown that this model exhibits chaos. See "Spiral chaos in a predator-prey model" by M. E. Gilpin, The American Naturalist, 1979, pages 306-308, and Chapter 7.

14. Suppose a model of two detritus compartments, a detritivore compartment, and a predator compartment is given by

$$\dot{x}_1 = x_1(-.1x_1 - 2x_3) + 1 + \sin 2\pi(t/365)$$
$$\dot{x}_2 = x_2(-2x_2 - 2x_3) + 1 + \sin 2\pi(t/365)$$
$$\dot{x}_3 = x_3(x_1 + x_2 - x_4)$$
$$\dot{x}_4 = x_4(.1x_3 - .001x_4)$$

where time is measured in days since January 1. Identify the compartments. What does the Trajectory Trapping Theorem say about this model?

15. Show $x(t) = (\sin(t), \cos(t))$ is a trajectory for

$$\dot{x}_1 = x_2$$
$$\dot{x}_2 = -x_1(2x_1^2 + 2x_2^2 - 1)$$

What is the period of this cyclic trajectory? Find three constant trajectories for this system.

16. Calculate the linear approximation matrix A (community matrix) of the system

$$\dot{x}_1 = .1x_1^2 - 2x_1x_2 + 2.1x_1$$
$$\dot{x}_2 = 2x_2x_1 - .1x_2^2 - 1.9x_2$$

about the constant trajectory $\hat{x}(t) = (1,1)$. Use the Hurwitz test and the Linearization Theorem to show that $(1,1)$ is an attractor trajectory for the full system.

17.   Find the linear approximation matrix A (community matrix) for the dynamical system

$$\dot{x}_1 = x_1(1\text{-}x_2)$$
$$\dot{x}_2 = x_2(x_1\text{-}x_3)$$
$$\dot{x}_3 = x_3(2x_2\text{-}x_3\text{-}x_4)$$
$$\dot{x}_4 = x_4(x_3\text{-}x_5)$$
$$\dot{x}_5 = x_5(x_4\text{-}1)$$

about the constant trajectory (1,1,1,1,1). This model might be thought of as a simplistic version of four predation interactions. What would be the predator-prey relationships?

18. Show the linear approximation matrix A(t) for the dynamical system

$$\dot{x}_1 = \text{-}x_2\text{-}x_1+4+\cos(t)$$
$$\dot{x}_2 = x_1\text{-}2$$

about the trajectory $\hat{x}(t) = (\cos(t)+2, \sin(t)+2)$ is

$$\begin{pmatrix} \text{-}1 & \text{-}1 \\ 1 & 0 \end{pmatrix}$$

What do the Hurwitz test and the Linearization Theorem imply?

19.   Show the linear approximation matrix A(t) for

$$\dot{x}_1 = \cos(t)+(4+2\sin(t))x_1\text{-}x_1^2\text{-}x_1x_3$$
$$\dot{x}_2 = \cos(t)+(4+2\sin(t))x_2\text{-}x_2^2\text{-}x_2x_3$$
$$\dot{x}_3 = \cos(t)+x_1x_3+x_2x_3\text{-}2x_3^2$$

about the trajectory $\hat{x}(t) = (2+\sin(t), 2+\sin(t), 2+\sin(t))$ is

$$A(t) = \begin{pmatrix} -2-\sin(t) & 0 & -2-\sin(t) \\ 0 & -2-\sin(t) & -2-\sin(t) \\ +2+\sin(t) & +2+\sin(t) & -4-2\sin(t) \end{pmatrix}$$

Suppose $y(t)$ is a typical solution of the linear dynamical system with time dependent coefficients $\dot{y} = A(t)y$. Calculate the rate of change along $y(t)$ of the function Λ defined by $\Lambda = y_1^2 + y_2^2 + y_3^2$. Explain whether or not the Hurwitz test can be applied to this problem. Can Λ be used as a Lyapunov function? Can the Linearization Theorem be applied?

20. Consider the Lorenz dynamical system

$$\dot{x}_1 = -10x_1 + 10x_2$$
$$\dot{x}_2 = -x_1 x_3 + 28x_1 - x_2$$
$$\dot{x}_3 = x_1 x_2 - (8/3)x_3$$

Use the Hurwitz test to comment on the stability of this model about the constant trajectory $\hat{x}(t) = (0,0,0)$.

21.   Consider the system

$$\dot{x}_1 = x_1(2 - x_1 - x_2)$$
$$\dot{x}_2 = x_2(x_1 - x_2)$$

Find three constant trajectories in the nonnegative orthant. Try to apply the Hurwitz test and the Linearization Theorem to the linear approximation of this system at each constant trajectory.

ANSWERS TO PROBLEMS

1. One constant trajectory is clearly $(0,0)$. If $x_1 \neq 0$, then $x_2 = \pm 1$, so $x_1 = \pm 1$. Thus the other constant trajectories are $(+1,+1)$, $(+1,-1)$, $(-1,+1)$, and $(-1,-1)$.

2. $\dot{x}_1 = 0$ if and only if $x_2 = -x_1$; such points lie on the straight line with slope -1 passing through $(0,0)$. $\dot{x}_2 = 0$ if and only if $x_2 = x_1$; such points lie on the straight line with slope $+1$ passing through $(0,0)$.

3. $\dot{x}_1 = te^{-t} \cdot \frac{1}{2} t^2 e^{-t} + e^{-t} - te^{-t} - e^{-t} = -x_1 + x_2$; $\dot{x}_2 = e^{-t} - te^{-t} - e^{-t} = -x_2 + x_3$; and $\dot{x}_3 = -e^{-t} = -x_3$.

4. So long as $x(t) > K$, $d(x(t),K) = x(t) - K$. Thus $(x\dot(t)-K) = rK^{-1} x(K-x) < 0$.

5. We observe: $f(K) = 0$; $f'(x) = 1/K - 1/x$, so $f'(x) = 0$ only at $x = K$; and $f''(x) = x^{-2}$, so $f''(x)$ is positive for all positive x, in particular for $x = K$. Also
$$\lim_{x \to \infty} f'(x) = K^{-1},$$
so $f(x)$ grows approximately linearly as $x \to \infty$.
On the other hand, $\lim_{x \to \infty} f'(x) = \lim_{x \to \infty} -\ln(x/K) - 1 = +\infty$. Hence the graph
of $f(x)$ for positive x is U-shaped, has positive values except at $x = K$, and is unbounded as $x \to 0$ and as $x \to +\infty$.

6. Considering the results of the previous problem, the level sets of $\Lambda = x/K -\ln(x/K) - 1$ are pairs of points, one point between 0 and K, the other $> K$. For any such level set there exists a pair of points equidistant from K and closer to K than either level set point. Also, for any pair of points equidistant from K, there exists a level set pair of points, each closer to K than the equidistant points. The rate of change of Λ along a trajectory $x(t)$ is

$$\dot{\Lambda} = \dot{x}/K - \dot{x}/x = -rK^{-2}(x-K)^2.$$

Thus $\dot{\Lambda}$ is negative and bounded away from zero for all x beyond any given distance from K. These qualities can be used as in section 3 of this chapter to establish that $x(t) = K$ is an attractor trajectory with basin of attraction being the entire positive orthant, that is, all $x > 0$.

7. $\partial f/\partial x_2 = x_1 x_3$; at $(1,1,1,0)$ this function has the value 1. $\partial^2 f/\partial x_1 \partial x_3 = x_2$; at $(0,1,0,\pi)$ this function has the value 1. $\partial f/\partial t = -x_3\sin(t)$; at $(0,1,1,\pi/2)$ this function has the value -1.

8. $f'(x) = -1+2x-3x^2$, so $f'(2) = -9$. Thus points on the tangent line satisfy $y = -9x+b$. Since the tangent line passes through $(2,-5)$, $b = 13$.

9. The linear approximation is $0+(+3)(x_1-1)+(-3)(x_2-1) = 3x_1-3x_2$. At $(1.1,1.1)$ both f and the linear approximation of f have the value zero.

10. The linear approximation is $0+(+1)(x_1-0)+(0)(x_2-0+(0)(x_3-1+(-1)(t-\pi/2) = x_1-t+\pi/2$. At $(.1,.1,1.1,\pi/2+.1)$ both f and the linear approximation of f have the value zero.

11. The linear approximation is $0+(+1)(x_1-0)+(+1)(x_2-1)+(0)(x_3-0) = x_1+x_2-1$. At $(.1,1.1,.1)$ the linear approximation has value .2000 while the function itself has value .2098. At $(.01,1.01,.01)$ the linear approximation has value .0200 while the function itself has value .0201.

12.   Compartment 3 is the herbivore compartment; the predation interaction terms are $-2x_1x_3$ in $\dot{x}_1$, $-2x_2x_3$ in $\dot{x}_2$, and $x_3(x_1+1.2x_2)$ in $\dot{x}_3$. The trajectory Trapping Theorem implies that no trajectory for this model which starts in the positive orthant can reach the finite boundary of the positive orthant in a finite time interval.

13. Compartment 3 is the herbivore compartment. The Trajectory Trapping Theorem implies that no trajectory for this model which starts in the positive orthant can reach the finite boundary of the positive orthant after a finite time interval.

14. Compartments 1 and 2 are detritus compartments with detritus donation terms $1+\sin 2\pi(t/365)$ each. Compartment 3 is a detritivore compartment and compartment 4 is a predator compartment. The Trajectory Trapping Theorem implies that no trajectory for this model which starts in the positive orthant can reach the boundary of the positive orthant after a finite time interval.

15. $\dot{x}_1 = \cos(t) = x_2$ and $\dot{x}_2 = -\sin(t) = -\sin(t)\left[2\sin^2(t)+2\cos^2(t)-1 \right] =$

$-x_1(2x_1^2+2x_2^2-1)$. The period of the trajectory is 2π. Clearly $(0,0)$ is a constant trajectory. Any constant trajectory has $x_2 = 0$, so $x_1 \neq 0$ implies $x_1 = \pm (2^{1/2})/2$. Thus the three constant trajectories are $(0,0)$, $(+(2^{1/2})/2,0)$ and $(-(2^{1/2})/2,0)$.

16. The linear approximation matrix is

$$\begin{pmatrix} -.2x_1-2x_2+2.1 & -2x_1 \\ 2x_1 & 2x_1-.2x_2-1.9 \end{pmatrix}$$

which at $(1,1)$ has the value

$$\begin{pmatrix} -.1 & -2 \\ 2 & -.1 \end{pmatrix}$$

The characteristic polynomial of this matrix is $z^2+.2z+4.01$. The Hurwitz determinants are $.2$ and $(.2)(4.01)-(0)(1)$, both positive. Since the linear approximation matrix at $(1,1)$ is a matrix of constants, the Linearization Theorem can certainly be applied. Thus $(1,1)$ is an attractor trajectory for the full system.

17. The linear approximation matrix is

$$\begin{pmatrix} 1-x_2 & -x_1 & 0 & 0 & 0 \\ x_2 & x_1-x_3 & -x_2 & 0 & 0 \\ 0 & 2x_3 & 2x_2-2x_3-x_4 & -x_3 & 0 \\ 0 & 0 & x_4 & x_3-x_5 & -x_4 \\ 0 & 0 & 0 & x_5 & x_4-1 \end{pmatrix}$$

which at $(1,1,1,1,1)$ has the value

$$\begin{pmatrix} 0 & -1 & 0 & 0 & 0 \\ 1 & 0 & -1 & 0 & 0 \\ 0 & 2 & -1 & -1 & 0 \\ 0 & 0 & 1 & 0 & -1 \\ 0 & 0 & 0 & 1 & 0 \end{pmatrix}$$

In this model 5 preys upon 4, 4 preys upon 3, 3 preys upon 2, and 2 preys upon 1. Also 1 is an autotroph, 3 is auto-regulating, and 5 looses energy spontaneously.

18. That $\hat{x}(t)$ is a trajectory for the system follows from $\dot{x}_1 = -\sin(t) = -\sin(t)-2-\cos(t)-2+4+\cos(t) = -x_2-x_1+4+\cos(t)$, and $\dot{x}_2 = \cos(t) = \cos(t)+2-2 = x_1-2$. The linear approximation of the system about any trajectory is

$$\begin{pmatrix} -1 & -1 \\ 1 & 0 \end{pmatrix}$$

The characteristic polynomial is therefore z^2+z+1. The Hurwitz determinants are both 1. Since the linear approximation matrix is constant, the Linearization Theorem can be applied to show that $\hat{x}(t)$ is an attractor trajectory. In fact, any trajectory for the full system is an attractor trajectory.

19. Calculus can be used to show $\hat{x}(t)$ is indeed a trajectory for the system and that the linear approximation matrix is as shown. Since $A(t)$ is not a matrix of constants, the Hurwitz test cannot be applied. However, the behavior of $\Lambda = y_1^2+y_2^2+y_3^2$ along a trajectory of the linear system with time dependent coefficients is

$$\dot{\Lambda} = 2y_1\dot{y}_1+2y_2\dot{y}_2+2y_3\dot{y}_3$$

$$= +2y_1\left[-(2+\sin(t))y_1-(2+\sin(t))y_3\right]$$

$$+2y_2\left[-(2+\sin(t))y_2-(2+\sin(t))y_3\right]$$

$$+2y_3\left[(2+\sin(t))y_1+(2+\sin(t))y_2-(2+\sin(t))y_3\right]$$

$$= -2(2+\sin(t))y_1^2-2(2+\sin(t))y_2^2-4(2+\sin(t))y_3^2$$

$$\leq -2(y_1^2+y_2^2+y_3^2)$$

Of course, the level sets of Λ are spheres concentrically arranged about $0 = (0,0,0)$. Using Lyapunov arguments like those for predator-prey model one can show that 0 must be an attractor trajectory for the linearized system $\dot{y} = A(t)y$. Since the system functions of the original system are analytic, since the entries in $A(t)$ are bounded, and since 0 is an attractor trajectory for the $\dot{y} = A(t)y$, the Linearization Theorem implies $\hat{x}(t)$ is an attractor trajectory for the full system.

20.   The linear approximation matrix is

$$\begin{pmatrix} -10 & 10 & 0 \\ -x_3+28 & -1 & -x_1 \\ x_2 & x_1 & -8/3 \end{pmatrix}$$

About the constant trajectory $(0,0,0)$ we have

$$\begin{pmatrix} -10 & 10 & 0 \\ 28 & -1 & 0 \\ 0 & 0 & -8/3 \end{pmatrix}$$

The characteristic polynomial is $(z^2+11z-270)(z+8/3)$, the roots of which are $-8/3$ and (approximately) -22.83 and about 11.83. The existence of a positive root implies that $(0,0,0)$ is not an attractor trajectory for the full system.

    There are exactly two other constant trajectories for this model, both not attractor trajectories. In his important paper "Deterministic Nonperiodic Flow," Journal of the Atmospheric Sciences, Volume 20, 1963, pages 130-141, Edward Lorenz first established the existence of a new type of stability now called chaos. Chaos will be developed further in Chapter 7.

21. The linear approximation matrix here is

$$\begin{pmatrix} 2-2x_1-x_2 & -x_1 \\ x_2 & x_1-2x_2 \end{pmatrix}$$

Clearly (0,0) is a constant trajectory. We may find other constant trajectories as follows. If $x_2 \neq 0$, then $x_1 = x_2$ and $x_1+x_2 = 2$ imply $x_1 = x_2 = 1$. If $x_2 = 0$ and $x_1 \neq 0$, then $x_1 = 2$. About (0,0) the linear approximation matrix is

$$\begin{pmatrix} 2 & 0 \\ 0 & 0 \end{pmatrix}$$

The characteristic polynomial of this matrix is z^2-2z, the roots of which are 0 and 2. Since one root is positive, (0,0) cannot be an attractor trajectory.

About (2,0) the linear approximation matrix is

$$\begin{pmatrix} -2 & -2 \\ 0 & 2 \end{pmatrix}$$

The characteristic polynomial is z^2-4, the roots of which are +2 and -2. Since one root is positive, (2,0) cannot be an attractor trajectory.

About (1,1) the linear approximation matrix is

$$\begin{pmatrix} -1 & -1 \\ 1 & -1 \end{pmatrix}$$

The characteristic polynomial is z^2+2z+2 and the associated Hurwitz determinants are 2 and 4. Thus applying the Hurwitz test and the Linearization Theorem leads to the conclusion that (1,1) is an attractor trajectory for the full system.

CHAPTER FIVE
INTRODUCTION TO ECOSYSTEM MODELS

5.1 The Community Matrix

The idea of a community matrix was introduced in 1968 by R. Levins [L]. If a model has n compartments, then the community matrix is an n×n matrix of functions of time A(t). As in the previous chapter, A(t) is defined relative to a special trajectory $\hat{x}(t)$, a trajectory of observed, "normal," or, on the basis of experience, predicted compartment values. Recall the definition of A(t):

$$A_{ij}(t) = \frac{\partial f_i}{\partial x_j}(\hat{x}(t), t)$$

where $\hat{x}(t)$ is a trajectory of the ecosystem model (dynamical system)

$$\dot{x}_i = f_i(x,t)$$

In the very special and unrealistic case that A(t) is actually a matrix of constants, certain mathematical tests (the Hurwitz tests of the previous chapter) can be performed on A which indicate the qualitative nature of nearby trajectories, namely, whether or not relative to nearby trajectories $\hat{x}(t)$ is an attractor trajectory.

Let's now place the idea of a community matrix in its ecological context. Suppose the amount of energy (or other stuff which is conserved in the system, such as the number of carbon atoms) in compartment i is x_i. If at time t compartment i is "preyed upon" by compartment j (including browsing, parasitism, and all sorts of predation), then $A_{ij}(t) < 0$ and $A_{ji}(t) > 0$. It would be a little imprecise to state, "compartment j has a negative impact on compartment i," or "compartment i has a positive impact on compartment j." However, it is precisely true that if the system at time t is "on" $\hat{x}(t)$ and if the quantity x_j only is somehow instantaneously increased, then x_i will at least initially decrease. Likewise, an increase in x_i will cause at least initially an increase in x_j.

The entry $A_{ii}(t)$ is negative if compartment i is auto-regulating. If $A_{ii}(t) < 0$, if the system is on $\hat{x}(t)$, and if the quantity x_i only is somehow instantaneously increased, then x_i will at least initially decrease.

In animal compartments, auto-regulation can arise through intracompartmental predation and physical interference including territorial conflicts. In autotroph compartments, auto-regulation can arise from intracompartmental competition for light or nutrients. In detritus compartments, auto-regulation can arise from autolysis.

The following anecdotal examples should clarify the role of A(t).

Suppose a sparrow hawk rises to harass an overflying osprey, the osprey posing no threat to the sparrow hawk or her brood but resembling in silhouette other buteos which might pose a threat. Since both the osprey compartment and the sparrow hawk compartment lose energy in the encounter (as heat, ultimately), both entries in A(t) at the particular time t would be negative. That is, if the sparrow hawk compartment is compartment 1 and the osprey compartment is compartment 2, then at the time t of the encounter, $A_{12}(t)$ and $A_{21}(t)$ would be negative numbers.

Suppose next that both cattle and sheep graze a summer pasture and that the energy flow in the pasture is given by a model $\dot{x} = f(x,t)$ with observed trajectory $\hat{x}(t)$. If the pasture were rich, then both entries in A(t) corresponding to interactions between the cattle compartment and the sheep compartment would be zero.

Suppose the pasture is overgrazed. The cattle and sheep would then compete and the corresponding entries in A(t) would be negative. The competition terms in A(t) would not be due to removal of energy from the pasture by one compartment which might have been used by the other; such an effect would be reflected in $\hat{x}(t)$ values. Rather, a negative pair of entries in A(t) would arise if each compartment physically interferes with the grazing habits (the energy uptake rate) of the other. Thus the energy flow viewpoint results in conclusions about interactions which are not entirely the same as intuitive "positive effects" or "negative effects."

It is also possible that for some pair of compartments 1 and 2 that both A_{12} and $A_{21}(t)$ are positive. Such symbiotic compartments do not generally pass energy both ways between themselves, but they mutually facilitate energy uptake from or reduce energy loss to other compartments. For example, it is well known that flagella-bearing protozoans of the genus *Trichonympha* break down cellulose in the guts of termites and orthopterans such as cockroaches. However, it seems unlikely that at the ecosystem level of organization *Trichonympha* spp. and their hosts would be considered as separate compartments.

The example of algal and fungal cells living together to form a lichen is another famous case of symbiosis. Of course, the fungal cells actually prey upon (derive energy from) the algal cells, although algal photosynthesis is made possible in the first place by nutrients released from a substrate by the fungal cells. Again the example is rather contrived; it is unlikely that at the ecosystem level of organization lichens would be so divided.

It has been asserted by Robert M. May [M, p. 73] that the above "- -" and "+ +" pairs of entries in a community matrix are somewhat exceptional. Aside from competition between autotrophs and auto-regulation in general, the overwhelming majority of community matrix entries at any particular time seem to correspond to predation interactions (giving rise to "+ -" pairs of entries) or detritus donation (giving rise to "+ 0" pairs).

We have already mentioned some examples of predation interactions. Detritus donation refers, of course, to energy transfer which occurs at a rate independent of the recipient compartment. Suppose a tree growing at the edge of a lake is blown over in a storm and becomes a partly submerged log. Such an event would mark detritus donation from the tree compartment to the partly submerged log compartment. The log would thereafter serve as a base compartment for a miniature ecosystem, somewhat in parallel to its previous terrestrial role. This case is, of course, but one example of the general rain of detritus from plants onto the earth.

5.2 Predator-Prey Equations and Generalizations Thereof

In the previous two chapters we studied predator-prey differential equations of the type

$$\dot{x}_1 = x_1(a_{11}x_1 + a_{12}x_2 + b_1)$$
$$\dot{x}_2 = x_2(a_{21}x_1 + a_{22}x_2 + b_2)$$

In particular, trajectories for an analogous difference equation system with $a_{11} = -.1$; $a_{12} = -2$; $a_{21} = 2$; $a_{22} = -.1$; $b_1 = 2.1$; $b_2 = -1.9$ were calculated with various Δt values.

The classical n-dimensional extension of the predator-prey model (often called a *Lotka-Volterra model*) is

$$\dot{x}_i = x_i(\sum_{j=1}^{n} a_{ij}x_j + b_i) + \sum_{j=1}^{n} c_{ij}x_j$$

Thus for $i \neq j$, the term $x_i(a_{ij}x_j)$ is the gain (>0) or loss (<0) to compartment i by virtue of predation interactions with compartment j; of course, if compartments i and j are not involved in predation interactions, then $a_{ij} = 0$. The term $x_i(a_{ii}x_i)$ is presumed nonpositive and amounts to energy loss through auto-regulation. The term $x_i(b_i)$ is the rate of energy gain (>0) to autotroph compartment i in the absence of competition; or the rate of spontaneous energy loss (<0) from a nonautotroph compartment. If compartment i is a detritus compartment, then the term $c_{ij}x_j$ is the rate of detritus donation from compartment k to compartment i (possibly zero for some k); otherwise, all $c_{ij} = 0$.

This generalized predation and detritus donation model is obviously an extreme simplification of nature. However, some insight into how realistic models might be built can be gained by studying simple models.

A constant trajectory $\hat{x}(t) = \hat{x}$ for our n–dimensional model amounts to the existence of n positive numbers $\hat{x}_1, \hat{x}_2,\ldots, \hat{x}_n$ satisfying

$$0 = \hat{x}_i \left(\sum_{j=1}^{n} a_{ij}\hat{x}_j + b_i \right) + \sum_{j=1}^{n} c_{ij}\hat{x}_j$$

Let us suppose that such a constant trajectory exists and proceed to evaluate its stability. The entry A_{ij} in the linear approximation matrix on the constant trajectory $\hat{x}$ is

$$\delta_{ij} \left(\sum_{j=1}^{n} a_{ij}\hat{x}_j + b_i \right) + \hat{x}_i a_{ij} + c_{ij}$$

where $\delta_{ij} = 1$ if $i = j$ and $\delta_{ij} = 0$ if $j \neq j$ (δ_{ij} is the *Kronecker delta* symbol). Thus $i \neq j$ implies $A_{ij} = \hat{x}_i a_{ij} + c_{ij}$. Also, $A_{ii} = -\sum_{j=1}^{n} a_{ij}\hat{x}_j + b_i + \hat{x}_i a_{ii}$. Since $\hat{x}$ is a constant trajectory for the system, $\sum_{j=1}^{n} a_{ij}\hat{x}_j + b_i = -\sum_{j=1}^{n} c_{ij}\hat{x}_j/\hat{x}_i$, so A_{ii} can also be written as $-\sum_{j=1}^{n} c_{ij}\hat{x}_j/\hat{x}_i + \hat{x}_i a_{ii}$. Thus A_{ii} is always a nonpositive number (the sum of nonpositive numbers). In the event that the model has no detritus compartments (so all $c_{ij} = 0$), each sum $\sum_{j=1}^{n} a_{ij}\hat{x}_j + b_i = 0$ and each $A_{ij} = \hat{x}_i a_{ij}$.

Suppose a certain model of this type has six compartments and

$$a = \begin{pmatrix} -.1 & -.01 & -1 & 0 & 0 & 0 \\ 0 & -.2 & -1 & 0 & 0 & 0 \\ .1 & .1 & -1.9 & 0 & 0 & 0 \\ 0 & 0 & 0 & -1 & 0 & 0 \\ 0 & 0 & 0 & 0 & -1 & 0 \\ 0 & 0 & 0 & 0 & 0 & -.1 \end{pmatrix} \qquad b = \begin{pmatrix} 21 \\ 30 \\ -1 \\ -9 \\ -9 \\ -.9 \end{pmatrix}$$

$$c = \begin{pmatrix} 0 & 0 & 0 & 0 & 0 & 0 \\ 0 & 0 & 0 & 0 & 0 & 0 \\ 0 & 0 & 0 & 0 & 0 & 0 \\ 0.1 & 0 & 0 & 0 & 0 & 0 \\ 0 & 0.1 & 0 & 0 & 0 & 0 \\ 0 & 0 & 0.1 & 0 & 0 & 0 \end{pmatrix}$$

The reader should be able to verify that $\hat{x} = (100,100,10,1,1,1)$ is a constant trajectory for this system. In this hypothetical model compartments 1 and 2 represent autotrophs (since b_1 and b_2 are positive). Compartments 1 and 2 are preyed upon by compartment 3 (since $a_{13} < 0$, $a_{31} > 0$, $a_{23} < 0$, and $a_{32} > 0$). In some way compartment 2 crowds compartment 1 (since $a_{12} < 0$). Compartments 1, 2, and 3 donate detritus to compartments 4, 5, and 6, respectively (since $c_{41} > 0$, $c_{52} > 0$, and $c_{63} > 0$). All compartments are auto-regulating (since each $a_{ii} < 0$). Energy enters the system in compartments 1 and 2, and energy leaves the system in ways other than auto-regulation from compartments 3, 4, 5 and 6 (since b_3, b_4, b_5, and b_6 are negative).

We can check the stability of this system around $\hat{x} = (100,100,10,1,1,1)$ as follows. The community matrix is

$$A = \begin{pmatrix} -10 & -1 & -100 & 0 & 0 & 0 \\ 0 & -20 & -100 & 0 & 0 & 0 \\ 1 & 1 & -19 & 0 & 0 & 0 \\ .1 & 0 & 0 & -11 & 0 & 0 \\ 0 & .1 & 0 & 0 & -11 & 0 \\ 0 & 0 & .1 & 0 & 0 & -1.1 \end{pmatrix}$$

The characteristic polynomial of A is $(z^3+49z^2+970z+6700) \cdot (z+11)^2 \cdot (z+1.1)$. Clearly three of the roots are -11, -11, and -1.1, all negative real numbers. Therefore all the roots of the characteristic polynomial of A will have negative real parts if and only if all the roots of $z^3 + 49z^2 + 970z + 6700$ have negative real parts. So let us apply the Hurwitz test to the smaller polynomial. The three determinants are 49; 40,830; and 40,830·6700, all positive. From the Linearization Theorem we can then conclude that the constant trajectory $\hat{x}$ is an attractor trajectory.

5.3 Signed Digraphs

We have seen that ecosystem interactions near a certain trajectory $\hat{x}(t)$ can be thought of in terms of a community matrix. The purpose of this section is to think of a community matrix as a combinatorial object called a *signed digraph*. Such a signed digraph is a formalization of the ecological concept of a food web diagram.

Suppose we are given an n×n matrix A of real numbers, such as a community matrix at a particular time. The signed digraph associated with A consists of n labelled vertices (points in a plane) together with from 0 to n^2 edges, each edge beginning and ending at a vertex. Furthermore, each edge has an orientation (arrow) and sign (+ or -).The signed edge corresponds to a nonzero matrix entry by the following simple rule: $A_{ij} \neq 0$ if and only if there is an edge from vertex j to vertex i, signed as the sign of A_{ij}. The signed digraph corresponding to the 3×3 matrix

$$\begin{pmatrix} -1 & -2 & 0 \\ 1 & 0 & -2 \\ 0 & 3 & 0 \end{pmatrix}$$

is shown in Fig. 5.1.

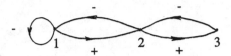

Figure 5.1. A signed digraph with three vertices.

We shall denote the signed digraph of A by SD(A). Predation interactions appear in signed digraph as pairs of edges, a "+" edge from prey to predator and a "-" edge from predator to prey. Detritus donation appears as one edge between two vertices, signed "+". Competition appears in a signed digraph as a "-" edge which is not part of a "+,-" 2-cycle.

A compartment in the n-dimensional model which is auto-regulating corresponds to a vertex in the signed digraph with a looping edge from the vertex to itself, signed "-". Considering the calculation in the previous section of A_{ii} for a detritus compartment, we note that a "-" looping edge is attached to every corresponding detritus vertex, whether or not the compartment is also auto-regulating.

The six-dimensional example used in the previous section thus corresponds to the signed digraph shown in Fig. 5.2.

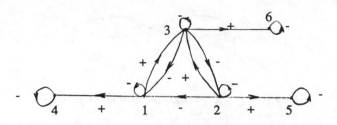

Figure 5.2. A signed digraph with six vertices.

An important concept in the study of signed digraphs is the p-cycle. Suppose by following consecutively p edges one can pass through p distinct vertices, returning to the initial vertex as the final step. Such a set of p edges and p vertices is called a *p-cycle*. In Fig. 5.2 we see six 1-cycles, two 2-cycles, one 3-cycle, and zero 4-cycles, 5-cycles, and 6-cycles.

5.4 Qualitative Stability of Linear Systems

Suppose 0 is an attractor trajectory for a linear system which is the linear approximation of a nonlinear system about a trajectory $\hat{x}(t)$. If the nonlinear system meets the criteria of the Linearization Theorem, then $\hat{x}(t)$ is an attractor trajectory for the nonlinear system. The Hurwitz test is a <u>quantitative</u> test for determining that 0 is an attractor trajectory for the linear system. In this section we shall study a <u>qualitative</u> test for determining that 0 is an attractor trajectory for the linear system. For the mathematical ecology background of these ideas, see [M, J]. A mathematical account of qualitative stability theory of linear dynamical system can be found in [JKvdD].

It is reasonable to suppose that determining the signs (+, -, or 0) of entries in a community matrix associated with an ecosystem model and trajectory $\hat{x}(t)$ is generally far easier than determining, that is, accurately estimating, the actual numerical values of the entries. Hence it is reasonable to ask: for what sign patterns of a

constant community matrix A is 0 necessarily an attractor trajectory for $\dot{x} = Ax$?
Any matrix or signed digraph associated with such a sign pattern is called *sign stable*. Somewhat surprisingly, there is a fairly rich variety of sign stable matrices.

As an example of a sign stable system, consider the decomposition of detritus in a woodland stream, decomposition due to mechanical and chemical processes as well as colonization by bacteria and fungi. We might have five compartments in a preliminary model. Suppose detritus is supplied from outside the system to compartment 1, so compartment 1 is in effect an autotroph, and suppose that otherwise detritus flows among the compartments as in the equations

$$\dot{x}_1 = 10^6 - x_1^2$$
$$\dot{x}_2 = .1x_1 - x_2$$
$$\dot{x}_3 = .1x_1 - x_3$$
$$\dot{x}_4 = .5x_3 + .5x_2 - x_4^2$$
$$\dot{x}_5 = .01x_3 - x_5^2$$

The digraph of the associated community matrix about the constant trajectory (1000,100,100,10,1) is shown in Fig. 5.3. The community matrix itself is

$$A = \begin{pmatrix} -1000 & 0 & 0 & 0 & 0 \\ .1 & -1 & 0 & 0 & 0 \\ .1 & 0 & -1 & 0 & 0 \\ 0 & .5 & .5 & -20 & 0 \\ 0 & 0 & .01 & 0 & -2 \end{pmatrix}$$

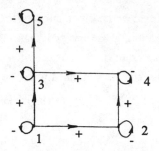

Figure 5.3. A community matrix digraph.

As we shall soon see, this community matrix is sign stable. In fact, any dynamical system derived from the above system by arbitrary changes in the magnitudes of the nonzero coefficients always has a constant trajectory in the positive orthant and the associated community matrix is always sign stable. Hence there is some qualitative aspect to the stability of a model of detritus donation when in the model detritus flows one-way through a hierarchy of compartments.

For a closely related realistic model of detritus flow with 39 compartments, see [B].

The conditions for sign stability of a matrix A can be couched in terms of the signed digraph SD(A) of A; particular importance is attached to the types of p-cycles in SD(A).

For A to be sign stable it is necessary that in SD(A) the following conditions are met:

α **any 1-cycle is signed "-";**

β **any 2-cycle is signed "+,-";**

γ **no p-cycles, p ≥ 3 occur.**

Conditions α, β, and γ already imply a rather rigid structure for SD(A). Within a signed digraph SD(A) which meets the conditions, the vertices may be thought of

as belonging to maximal blocks of vertices interconnected by 2-cycles. If a vertex i
is not part of any 2-cycle, then it is itself a maximal block. If two vertices are
connected by a 2-cycle, then they belong to the same maximal block. Furthermore,
condition γ implies that all edges connecting vertices in separate maximal blocks
should be directed one-way. We shall refer to such maximal blocks as predation
communities because in the context of ecosystem models, maximal blocks amount
to compartments interconnected by predation interactions. A representative signed
digraph with nine vertices, that is a representative community matrix for a model
with nine compartments, is shown in Fig. 5.4.

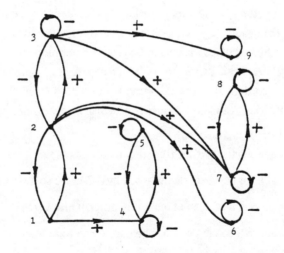

**Figure 5.4. A signed digraph associated with a nine compartment
community matrix.**

The signed digraph in Fig. 5.4 meets conditions α, β, and γ. It has five
predation communities. For illustrative purposes we might think of compartment 1

as an autotroph compartment which is preyed upon by compartment 2 and donates detritus to compartment 4, a plant litter compartment. In turn compartment 4 is preyed upon by aggregated detritivores in compartment 5. Top consumers in compartment 3 prey upon herbivores in compartment 2. Compartments 2 and 3 donate excreta to compartments 6 and 9. Compartments 2 and 3 also donate carrion to compartment 7. The detritus in compartment 7 is in turn consumed by detritivors in compartment 8. Compartments 3, 5, and 8 are auto-regulating. The 1-cycles attached to vertices 4, 6, 7, and 9 could result from auto-regulation (a_{ii} terms), from spontaneous loss terms (negative b_i terms), or from both types of terms.

In a complete description of sign stable matrices and signed digraphs it is necessary to specify which vertices have 1-cycles. It turns out that 1-cycles must occur in at least some key positions in SD(A) in order to guarantee sign stability, that is, that 0 will be an attractor trajectory for $\dot{x} = Ax$, regardless of the magnitudes of nonzero entries in A. This suggests that in a community matrix, stability hinges upon certain nondetritus compartments being auto-regulating or certain detritus compartments being auto-regulating or losing energy spontaneously.

Exactly which vertices must have 1-cycles is determined by two tests called color tests. Each test is an attempt to color all the vertices of a predation community black or white according to rules called Im rules or 0 rules ("zero rules"). A coloring of SD(A) according to the rules is called an *Im-coloring* or a *0-coloring*.

The Im rules are:

i.   **each vertex with a 1-cycle must be black;**

ii.   **each black vertex connected by a 2-cycle to a white vertex**

must be connected to at least one other white vertex;

iii. **each white vertex must be connected to at least one other**

white vertex by a 2-cycle.

The 0 rules are:

i. and ii., as in Im rules;

iii. **each white vertex must be connected to no other white**

vertex by a 2-cycle.

The reader should notice that coloring all vertices of SD(A) black is both an Im-coloring and a 0-coloring. Such colorings are called the *trivial colorings*. Any other coloring which exists for a given SD(A) is called a *nontrivial* coloring.

The reader should be able to show, one predation community at a time, that the signed digraph in Fig. 5.4 admits no nontrivial Im-coloring or 0-coloring.

The existence of a nontrivial Im-coloring of a signed digraph which satisfies α, β, and γ implies that some matrix $\tilde{A}$ of the given sign pattern exists with a purely imaginary eigenvalue. This is equivalent to the existence of a cyclic trajectory for $\dot{x} = \tilde{A}x$.

The existence of a nontrivial 0-coloring of a signed digraph which satisfies α, β, and γ implies that every matrix $\tilde{A}$ of the given sign pattern has zero as an eigenvalue. This is equivalent to the existence of a constant trajectory for $\dot{x} = \tilde{A}x$ which starts and stays arbitrarily close to 0; such a trajectory precludes the possibility that 0 is an attractor trajectory for $\dot{x} = \tilde{A}x$.

Consider the case of a simplistic model consisting of a single predation community given by

$$\dot{x}_i = x_i \left(\sum_{j=1}^{n} a_{ij} x_j + b_i \right)$$

with a constant trajectory for $\dot{x} = \tilde{A}x$. The linear approximation matrix of this system about $\hat{x}$ has in row i, column j, the number $\hat{x}_i a_{ij}$. It follows that the linear approximation matrix has the same entry signs as A itself. Thus the linear approximation matrix is sign stable if and only is A itself is sign stable. If A is sign stable and if the nonlinear system has a constant trajectory $\hat{x}$ in the positive orthant, then the Linearization Theorem and our sign stability observation imply that $\hat{x}$ is an attractor trajectory for the nonlinear system.

REFERENCES

[B] R. Boling, E. Goodman, J. Van Sickle, J. Zimmer, K. Cummins, R. Petersen, and S. Reice, Toward a model of detritus processing in a woodland stream, Ecology 56 (1975) 141-151.

[J] C. Jeffries, Qualitative stability and digraphs in model ecosystems, Ecology 55 (1974) 1415-1419.

[JKvdD] C. Jeffries, V. Klee, and P. van den Driessche, Qualitative stability of linear systems, Linear Algebra and Its Applications 87 (1987) 1-48.

[L] R. Levins, *Evolutions and Changing Environments* , Princeton U. Press, 1968.

[M] R. M. May, *Stability and Complexity in Model Ecosystems*, Princeton U. Press, 1973.

PROBLEMS

1. A hypothetical community matrix is given by

$$A = \begin{pmatrix} -2 & -1 & 0 \\ 2 & 0 & -1 \\ 0 & 3 & -2 \end{pmatrix}$$

Sketch the associated signed digraph SD(A) with three vertices and six edges. How many 1-cycles, 2-cycles and 3-cycles does SD(A) have?

2. A hypothetical community matrix is given by

$$A = \begin{pmatrix} -1 & 0 & 0 & -1 & 0 & 0 & 0 \\ 0 & 0 & 0 & -1 & -1 & 0 & 0 \\ 0 & 0 & 0 & 0 & -1 & 0 & 0 \\ +1 & +1 & 0 & 0 & 0 & -1 & 0 \\ 0 & +1 & +1 & 0 & 0 & -1 & 0 \\ 0 & 0 & 0 & +1 & +1 & -1 & -1 \\ 0 & 0 & 0 & 0 & 0 & +1 & -1 \end{pmatrix}$$

Sketch the associated signed digraph. Identify producers, first order consumers, second order consumers, and third order consumers. Is A sign stable?

3. A hypothetical community matrix is given by

$$A = \begin{pmatrix} -1 & 0 & -1 & 0 & 0 & 0 & 0 \\ 0 & 0 & -1 & 0 & 0 & 0 & 0 \\ +1 & +1 & 0 & -1 & 0 & 0 & 0 \\ 0 & 0 & +1 & -1 & 0 & 0 & 0 \\ +1 & +1 & 0 & 0 & 0 & -1 & 0 \\ 0 & 0 & 0 & 0 & +1 & 0 & 0 \\ 0 & 0 & 0 & +1 & 0 & 0 & -1 \end{pmatrix}$$

Sketch SD(A). Identify the compartments in the three predation communities of this model. Why is A not sign stable? Suppose we could attach one "-" 1-cycle to

SD(A), that is, replace some zero diagonal (A_{ii}) entry with a negative number. Where would we attach such a 1-cycle in order to make the corresponding matrix sign stable?

4. Show

$$A = \begin{pmatrix} 0 & -1 & 0 & 0 & 0 \\ +1 & 0 & -1 & 0 & 0 \\ 0 & +1 & -1 & -1 & 0 \\ 0 & 0 & +1 & 0 & -1 \\ 0 & 0 & 0 & +1 & 0 \end{pmatrix}$$

is not sign stable. How many additional negative diagonal entries in A would be needed to achieve sign stability?

5. Show

$$A = \begin{pmatrix} 0 & 0 & -1 & 0 & 0 \\ 0 & 0 & -1 & 0 & 0 \\ +1 & +1 & 0 & -1 & 0 \\ 0 & 0 & +1 & 0 & -1 \\ 0 & 0 & 0 & +1 & 0 \end{pmatrix}$$

has a nontrivial Im coloring and a nontrivial 0-coloring. If A_{44} and A_{55} are both changed to -1, is the new matrix sign stable?

6. Suppose in a three-dimensional predator-prey model (no detritus compartments) is given by

$$\dot{x}_1 = x_1(-2x_1 - x_2 + 10)$$
$$\dot{x}_2 = x_2(2x_1 - x_3)$$
$$\dot{x}_3 = x_3(3x_2 - 2x_3)$$

Find a constant trajectory for this model in the positive orthant. Use the Hurwitz test and the Linearization Theorem to show the constant trajectory is an attractor trajectory.

7. Use the sign stability test and the Linearization Theorem to show the constant trajectory for the system in problem 6 is an attractor trajectory.

8. Prove that any model obtained from the model in problem 6 by arbitrarily changing the magnitudes of nonzero coefficients in the model has a constant attractor trajectory in the positive orthant.

9. Consider the pattern of energy flow in a highly aggregated ecosystem model represented by the following diagram.

$$p \left(\begin{array}{l} \text{1. producer} \xrightarrow{\quad d \quad} \text{3. detritus} \qquad \text{6. detritus} \\ \\ \text{2. consumer} \xrightarrow{\quad d \quad} \text{4. detritus} \xrightarrow{\quad p \quad} \text{5. detritivore} \end{array} \right) d$$

Here edges marked "p" correspond to predation interactions with the obvious orientation and edges marked "d" correspond to detritus donation. Assuming each compartment is auto-regulating, derive the associated 6×6 matrix of signs (each entry +, -, or 0). Derive the associated signed digraph with six vertices and thirteen edges. Is the signed digraph sign stable?

10. Sketch two dissimilar signed digraphs each with four vertices, one "+,+,+,+" 4-cycle, no 3-cycles, two "+,-" 2-cycles, and zero 1-cycles.

11. Suppose a signed digraph SD(A) fulfills conditions α, β, and γ, and suppose a "-" 1-cycle is attached to every vertex. Is A sign stable?

12. Suppose a signed digraph SD(A) fulfills conditions α, β, and γ. A vertex in such a signed digraph is called peripheral if it is connected by a "+,-" 2-cycle to at most one other vertex. Suppose a "-" 1-cycle is attached to every peripheral vertex. Is A sign stable?

13. Consider the model $\dot{x}_i = x_i \left(\sum_{j=1}^{6} a_{ij} x_j + b_i \right)$ with

$$a = \begin{pmatrix} -1 & -1 & -1 & -1 & 0 & 0 \\ .1 & -1 & 0 & 0 & 0 & 0 \\ .1 & 0 & 0 & -1 & 0 & 0 \\ .1 & 0 & .1 & -1 & -.1 & 0 \\ 0 & 0 & 0 & .01 & 0 & -.1 \\ 0 & 0 & 0 & 0 & .01 & -.1 \end{pmatrix} \qquad b = \begin{pmatrix} 130 \\ 0 \\ 0 \\ 0 \\ 0 \\ 0 \end{pmatrix}$$

Show $\hat{x}(t) = (100,10,10,10,10,1)$ is a constant trajectory for this system. Compute the community matrix about $\hat{x}$. Sketch the associated signed digraph. Which vertex corresponds to a autotroph and which vertex corresponds to a top consumer?

14. The following diagram represents hypothetical energy flow in a lake ecosystem:

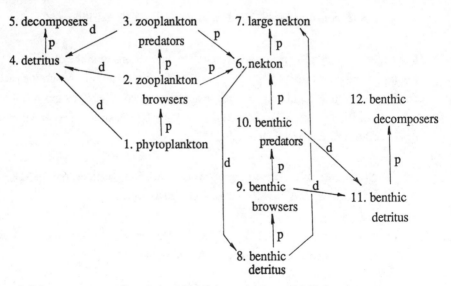

Here "p" denotes energy flow by predation interactions and "d" denotes energy flow by detritus donation. Compartments 4, 8, and 11 represent three types of detritus. Assuming the system can be modelled as some Lotka Volterra model

$$\dot{x}_i = x_i \left(\sum_{j=1}^{12} a_{ij}x_j + b_i \right) + \sum_{j=1}^{12} c_{ij}x_j$$

with each $a_{ii} < 0$, and assuming the model admits a constant trajectory in the positive orthant, sketch the associated signed digraph with 12 vertices, omitting "-"

1-cycles for the sake of clarity.

15. What does the trajectory Trapping Theorem imply about the Lotka-Volterra model

$$\dot{x}_i = x_i(\sum_{j=1}^{12} a_{ij}x_j + b_i) + \sum_{j=1}^{12} c_{ij}x_j$$

ANSWERS TO PROBLEMS

1. SD(A) contains two 1-cycles, two 2-cycles, and zero 3-cycles. A sketch of SD(A) is shown in Fig. 5.5.

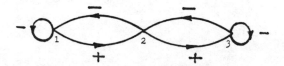

Figure 5.5. A signed digraph with three vertices.

2. A sketch of SD(A) is shown in Fig. 5.6 Compartments 1, 2, and 3 correspond to producers, compartments 4 and 5 correspond to first order consumers, compartment 6 corresponds to a second order consumer, and compartment 7 corresponds to a third order consumer. Since SD(A) contains two 4-cycles, A is not sign stable.

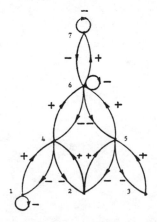

Figure 5.6. A signed digraph with two 4-cycles.

3. A sketch of SD(A) is shown in Fig. 5.7. Compartments 1, 2, 3, and 4 comprise one predation community; compartment 5 and 6 comprise a second predation community; and compartment 7 comprises a third predation community. Coloring vertices 5 and 6 white, all others black, is a nontrivial Im-coloring, so A is not sign stable. Sign stability would follow from attaching a "-" 1-cycle to vertex 5 or vertex 6.

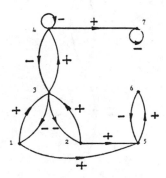

Figure 5.7. A signed digraph with three predation communities.

4. SD(A) in this case is a "straight chain" of four 2-cycles. Coloring the middle vertex black and the other vertices white is a nontrivial Im-coloring, so A is not sign stable. An additional "-" cycle would lead to sign stability.

5. A sketch of SD(A) is shown in Fig. 5.8. Coloring all vertices white is a nontrivial Im-coloring and coloring vertices 1 and 2 white, the others black, is a nontrivial 0-coloring. Adding "-" 1-cycles to vertices 4 and 4 would not preclude the 0-coloring.

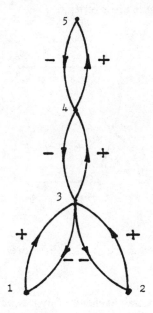

Figure 5.8. A signed digraph with five vertices.

6. A constant trajectory in the positive orthant has $\hat{x}_1 = \hat{x}_3/2$; $\hat{x}_2 = 2\hat{x}_3/3$; and $2\hat{x}_1 + \hat{x}_2 = 10$. Thus $\hat{x} = (3,4,6)$. The associated linear approximation matrix is

$$A = \begin{pmatrix} -6 & -3 & 0 \\ 8 & 0 & -4 \\ 0 & 18 & -12 \end{pmatrix}$$

The characteristic polynomial of A is $z^3 + 18z^2 + 168z + 720$. The determinants of the Hurwitz matrices are 18, 2304, and 2304·720; since all are positive, 0 is an attractor trajectory for the linear system $\dot{x} = Ax$. The Linearization Theorem can be used to show (3,4,6) is as attractor trajectory for the full nonlinear system.

7. The signed digraph of the linear approximation matrix in problem 6 is the same as SD(A) in Fig. 5.5. It can be shown that SD(A) is sign stable. Thus the Linearization Theorem can be used to show (3,4,6) is an attractor trajectory.

8. For any model of the type in problem 6

$$\dot{x}_1 = x_1 \, (-a_{11}x_1 - a_{12}x_2 + b_1)$$
$$\dot{x}_2 = x_2 \, (a_{21}x_1 - a_{23}x_3)$$
$$\dot{x}_3 = x_3 \, (a_{32}x_2 - a_{33}x_3)$$

with all "a" and "b" coefficients positive, a constant trajectory in the positive orthant is given by $\hat{x}_3 = b_1[a_{11}a_{23}a_{21}^{-1} + a_{12}a_{33}a_{32}]^{-1}$; $\hat{x}_1 = a_{23}a_{21}\hat{x}_3$; and $\hat{x}_2 = a_{33}a_{32}\hat{x}_3$. The signed digraph associated with this system and constant trajectory is again the same as in Fig. 5.5. The Linearization Theorem implies that the constant trajectory is an attractor trajectory.

9. The sign pattern of the linear approximation matrix (community matrix) of this system is

$$
\begin{pmatrix}
- & - & 0 & 0 & 0 & 0 \\
+ & - & 0 & 0 & 0 & 0 \\
+ & 0 & - & 0 & 0 & 0 \\
0 & + & 0 & - & - & 0 \\
0 & 0 & 0 & + & - & 0 \\
0 & 0 & 0 & 0 & + & -
\end{pmatrix}
$$

The associated signed digraph is shown in Fig. 5.9. The signed digraph is sign stable.

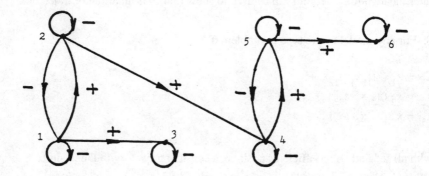

Figure 5.9. A signed digraph with six vertices.

10. Two signed digraphs which met the criteria of the problem are shown in Fig. 5.10.

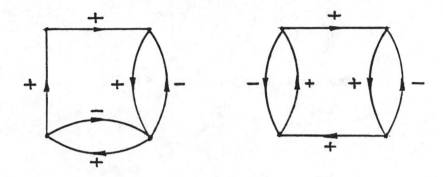

Figure 5.10.

11. Such a signed digraph can only be colored all black. It follows that such a signed digraph must be sign stable.

12. Such a signed digraph must be sign stable, as we now prove. Suppose there is a nontrivial Im-coloring or 0-coloring with at least one white vertex in at least one predation community. Since every peripheral vertex is black, at least one white vertex must be connected to a black vertex. That black vertex must be connected to at least one other white vertex. In turn, that white vertex or a vertex in the maximal block of white vertices containing it must be connected to another black vertex. Using condition γ we ultimately contradict the finiteness of n. This contradiction established sign stability.

13. The community matrix for this model about (100,10,10,10,10,1) is

126

$$A = \begin{pmatrix} -100 & -100 & -100 & -100 & 0 & 0 \\ 1 & -10 & 0 & 0 & 0 & 0 \\ 1 & 0 & 0 & -10 & 0 & 0 \\ 1 & 0 & 1 & -10 & -1 & 0 \\ 0 & 0 & 0 & .1 & 0 & -1 \\ 0 & 0 & 0 & 0 & .01 & -.1 \end{pmatrix}$$

The associated signed digraph is shown in Fig. 5.11.

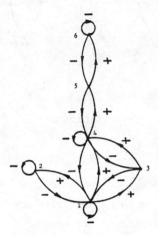

Figure 5.11. A signed digraph with six vertices. Vertex 1 corresponds to an autotroph and vertex 6 corresponds to a top consumer.

14. The signed digraph for this problem is shown in Fig. 5.12.

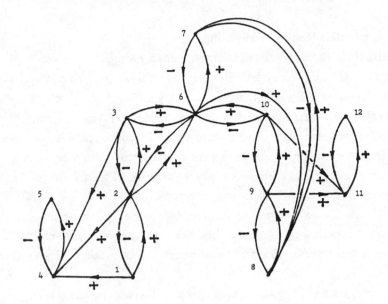

Fig. 5.12. A signed digraph corresponding to a hypothetical lake energy flow model.

15. The system functions of the Lotka-Volterra model may be written as $x_i g_i(x,t) + h_i(x,t)$ where $g_i(x,t) = \sum_{j=1}^{n} a_{ij} x_j + b_i$ and $h_i(x,t) = \sum_{j=1}^{n} c_{ij} x_j$. Note that

each $c_{ij} \geq 0$. Thus such a decomposition fulfills the criteria of the Trajectory Trapping Theorem. It follows that no trajectory of a Lotka-Volterra system which starts in the positive orthant can reach the finite boundary of the positive orthant in a finite time interval.

CHAPTER SIX
QUALITATIVE STABILITY OF ECOSYSTEM MODELS

6.1 Qualitative Results in Modeling

By virtue of the first law of thermodynamics, energy is conserved in ecosystems. Certainly ecosystem models based on energy flow should conserve energy. However, it is generally regarded as very difficult or impossible to specify exactly all the significant energy flow rates in an ecosystem, so every model inherently involves simplifications and approximations of nature. We know that the flow rates must "add up" so that energy is conserved; that is, for every compartment energy input minus energy output must exactly equal energy change. But we often do not know precisely the energy flow rates.

Fortunately there are mathematical results which amount to qualitative blueprints for building ecosystem models. Such results are the mathematical progeny of the qualitative theory of linear models in section four of Chapter Five, the theory of sign stability.

Let us assume throughout this chapter that all system functions in all ecosystem models considered have continuous first derivatives throughout the positive orthant of state space and for all time. Thus the Trajectory Existence Theorem implies that any initial trajectory value at any initial time can be extended uniquely in time as a trajectory of the model.

To start, let us consider the two-dimensional model in which herbivore compartment 2 preys upon autotroph compartment 1 as follows.

$$\dot{x}_1 = x_1 e_{12}(t) r_{12}(t) x_2 + p_1(t) x - a_1(t) x_1^2 - h_1(t) x_1 - 1_1(t) x_1$$

$$\dot{x}_2 = x_2 e_{21}(t) r_{21}(t) x_1 - a_2(t) x_2^2 - h_2(t) x_2 - 1_2(t) x_2$$

Thus the functions e, r, p, a, h, and 1 are all functions of time. The function $e_{12}(t)$ is the energy loss ($e_{12}(t) \leq 0$) to compartment 1 per predation interaction with compartment 2 at time t. The function $r_{12}(t)$ is the rate of predation interaction between compartments 1 and 2, per unit of compartment 1 energy, per unit of compartment 2 energy. Since $r_{12}(t)$ is defined analogously, $r_{21}(t) = r_{12}(t) \geq 0$. The function $p_1(t) x_1 - a_1(t) x_1^2$ is the net rate of energy change due to photosynthesis and competition. Auto-regulation effects such as intracompartmental competition are lumped in the a_1 term. Similarly $-a_2(t)x_2^2$ accounts for auto-regulation effects in compartment 2. The functions $-h_1(t)x_1$ and $-l_1(t)x_1$ represent energy loss from compartment 1 as heat or detritus through mechanisms not related to predation interactions or auto-regulation. Similarly $-h_2(t)x_2$ and $-l_2(t)x_2$ are defined.

Notice that the terms in the equation for $\dot{x}_1$ which are negative, that is, which correspond to the loss of energy from compartment 1, are of the form $x_1g_1(x,t)$ where $g_1(x,t)$ is defined and continuous for all x in the positive orthant and for all t. Analogous remarks can be made about $\dot{x}_2$. Thus the Trajectory Trapping Theorem can be applied to our model; we can conclude that no trajectory which starts in the positive orthant can reach the finite boundary of the positive orthant in finite time.

Of course, not every model of the above form has trajectories which are suggestive of predator-prey dynamics. There is too much artibrariness in the choice of system functions. Suppose we decide that the existence of an attractor trajectory would be a desirable model feature. Then it is natural to ask what additional mathematical conditions the model could fulfull in order to guarantee qualitatively the existence of an attractor trajectory. Such mathematical conditions might be reflected in nature, an application of the ancient game of deductive reasoning.

What follows is an account of such conditions which guarantee the existence of an attractor trajectory.

We shall need the ratio $e_{21}(t)/[-e_{12}(t)]$ of energy gain over energy lost at time t, a measure of *energy conversion efficiency*. Note that energy conversion efficiency is defined for any time when predation interactions occur and at such times it is a positive number.

Suppose there exists a trajectory $x(t) = (x_1(t), x_2(t))$ (not generally a constant trajectory) with two special properties. First, $\hat{x}(t)$ must be bounded as

$$0 < b_1 < \hat{x}_1(t) < b_2$$
$$0 < b_1 < \hat{x}_2(t) < b_2$$

where b_1 and b_2 are positive numbers. Second, for all times t when predation interactions occur, the ratio $\hat{x}_2(t)/\hat{x}_1(t)$ must be proportional to energy conversion efficiency, that is,

$$\hat{x}_2(t)/\hat{x}_1(t) = \kappa e_{21}(t)/[-e_{12}(t)]$$

for some positive constant κ. This proportionality means that in times of prey abundance (a season of year, for example) energy conversion efficiency is relatively low. If this phenomenon actually occurs in nature, energy conversion efficiency changes probably would be associated with behavioral changes, not biochemical changes.

We also require that the auto-regulation coefficients $a_1(t)$ and $a_2(t)$ be bounded below as

$$\alpha < a_1(t)$$
$$\alpha < a_2(t)$$

for some positive number α.

Given these assumptions on the trajectory $\hat{x}(t)$ and the auto-regulation coefficients $a_1(t)$ and $a_2(t)$, $\hat{x}(t)$ is actually an attractor trajectory, as we now prove.

Consider the function $\Lambda(x,t)$ defined in state space-time by

$$\Lambda(x,t) = \sum_{i=1}^{2} \lambda_i \left[\frac{x_i}{\hat{x}_i(t)} - 1 - \ln\left(\frac{x_i}{\hat{x}_i(t)} \right) \right]$$

where $\lambda_1 = 1$ and λ_2 satisfies $\lambda_2 e_{21}(t) r_{21}(t)\hat{x}_1(t) = -\lambda_1 e_{12}(t) r_{12}(t)\hat{x}_2(t)$ for all t. For any t with $r_{12}(t) = r_{21}(t) = 0$, the latter equation is trivial; otherwise λ_2 exists by virtue of the proportionality of $\hat{x}_2(t)/\hat{x}_1(t)$ and $e_{21}(t)/[-e_{12}(t)]$. A lengthy but straightforward calculation shows that along an arbitrary trajectory $x(t) = (x_1(t), x_2(t))$ in the positive orthant that

$$\dot{\Lambda}(x(t),t) = \sum_{i=1}^{2} -\lambda_i a_i(t) \frac{\left(x_i(t) - \hat{x}_i(t) \right)^2}{\hat{x}_i(t)}$$

Thus, as a function of state space-time

$$\dot{\Lambda}(x,t) = \sum_{i=1}^{2} -\lambda_i a_i(t) \frac{\left(x_i - \hat{x}_i(t)\right)^2}{\hat{x}_i(t)} < \sum_{i=1}^{2} -\frac{\alpha \lambda_i \left(x_i - \hat{x}_i(t)\right)^2}{b_2}$$

By making use of the boundedness of $\hat{x}_1(t)$ and $\hat{x}_2(t)$ one can show that conditions 1 and 2 of the Lyapunov Theorem are fulfilled. Outside a level cylinder of radius ε about $\hat{x}(t)$ and in the positive orthant, $\dot{\Lambda} \leq -\beta$ where β is a positive number smaller than either $\alpha\lambda_1\varepsilon^2/b_2$ or $\alpha\lambda_2\varepsilon^2/b_2$. This implies that condition 3 of the Lyapunov function is also fulfilled. Thus the Lyapunov Theorem can be used to prove that $\hat{x}(t)$ is an attractor trajectory. Since any point in state space-time is on some level set of Λ and since the above arguments are independent of position in state space-time, the basin of attraction of $\hat{x}(t)$ is all of the positive orthant and all of time.

6.2   Holistic Ecosystem Models

In this section we explore a fairly general type of ecosystem model. Two types of energy flow are allowed: predation interactions and detritus donation. Predation interaction energy flow depends upon the current state of the recipient compartment and ditritus donation energy flow does not. A predation community in this section means a set of compartments which are interconnected directly or indirectly by predation interactions.

Our holistic ecosystem model will generally have photosynthesis-based predation communities and detritus-based predation communities. Energy will be donated from upstream predation communities to downstream communities. We shall make use of a presumed hierarchy of such detritus donation relationships.

The model is built from the following types of pieces, coefficient functions of time. The general equation is

$$\dot{x}_i = \sum_{j=1}^{n} x_i e_{ij}(t) r_{ij}(t) x_j - a_i(t) x_i^2 + p_i(t) x_i - h_i(t) x_i + d_i(x,t) - l_i(t) x_i$$

where

$r_{ij}(t) x_i x_j$, $i \neq j$, is the rate of predation interactions between compartments i and j;

$e_{ij}(t)$, with $i \neq j$ and $r_{ij}(t) \neq 0$, is the rate of gain ($e_{ij}(t)>0$) or loss ($e_{ij}(t)<0$) to compartment i per predation interaction with compartment j;

$-a_i(t)x_i^2$ for nonautotroph compartments is the rate of energy loss from compartment i due to intracompartmental predation, territorial conflicts, autolysis, or other mechanisms not involving other compartments;

$p_i(t)x_i - a_i(t)x_i^2$ for autotroph compartments is the rate of potential photosynthesis gain minus intracompartmental competition loss, that is, net photosynthetic gain;

$-h_i(t)x_i$ is the rate of energy loss from compartment i to sinks outside the model including energy losses as heat, methane, or peat;

$d_i(x,t)$ for a detritus compartment is the rate of detritus gain from upstream compartments;

$-l_i(t)x_i$ is the rate of energy loss from compartment i to detritus compartments due to excretion, physical breakdown, or other nonpredation mechanisms.

Considering the role of each type of coefficient function, we assume each r, a, p, h, d, and l is nonnegative. This assumption and the Trajectory Trapping Theorem automatically imply that any trajectory starting in the positive orthant cannot reach the finite boundary of the positive orthant in finite time.

A strong mathematical assumption made in writing down this version of a holistic ecosystem model is that the coefficient functions r_{ij}, e_{ij}, a_i, p_i, h_i, and l_i (not $d_i(x,t)$) are all functions of time only.

A directed graph (not every edge is signed) which we shall call a food web can be assigned to this model at each time t as follows. The food web has n vertices. If $e_{ij}(t)r_{ij}(t) \neq 0$, then an edge is directed from vertex j to vertex i, signed as $e_{ij}(t)r_{ij}(t)$. Of course, such predation interaction edges in the food web occur in "+,-" 2-cycles. If $d_i(x,t)$ depends on compartment j in any way, then an unsigned edge is directed from vertex j to vertex i.

Thus associated with nonzero $e_{ij}(t)r_{ij}(t)$ entries are "+,-" 2-cycles in the food web. A maximal subset of vertices in a food web interconnected by such predation interaction 2-cycles is called a predation community. The mathematical analysis of the model is greatly simplified when detritus donation edges in the food web occur only between vertices in distinct predation communities. To be mathematically precise, the mathematical analysis is simplified when no p-cycle, $p \geq 0$, involves nodes in more than one predation community. We shall say that such a food web amounts to a hierarchy of predation communities with one-way energy flow in the hierarchy, or, to be brief, that such a food web has *one-way energy flow*.

Suppose in a food web with one-way energy flow that detritus donation edges are directed from some compartments in predation community I to some compartments in predation community II. We shall describe predation community I as *upstream* of predation community II and predation community II as *downstream* of predation community I.

Within a predation community the net energy conversion efficiency along a chain of predation interactions from one compartment at a low trophic level to another compartment at a high trophic level is just the product of the energy conversion efficiencies of the predation interactions in the chain. If the net energy conversion efficiency between two such compartments connected in the predation community by two or more such chains is at each time the same for each chain, the predation community is said to have *balanced predation loops*. Suppose in a predation community compartment 4 preys upon compartments 2 and 3, and compartments 2 and 3 prey upon compartment 1. If the predation community has balanced predation loops, then at each time t

$$\begin{bmatrix} e_{21}(t) \\ \overline{-e_{12}(t)} \end{bmatrix} \cdot \begin{bmatrix} e_{42}(t) \\ \overline{-e_{24}(t)} \end{bmatrix} = \begin{bmatrix} e_{31}(t) \\ \overline{-e_{13}(t)} \end{bmatrix} \cdot \begin{bmatrix} e_{43}(t) \\ \overline{-e_{34}(t)} \end{bmatrix}$$

6.3 Holistic Ecosystem Models with Attractor Trajectories

The purpose of this section is to present certain conditions on the coefficient functions of the holistic ecosystem model and to prove that any model which fulfills the conditions has an attractor trajectory. The results in this section are taken from [J].

Here are the conditions which we shall use in this section, expressed realtive to a given trajectory $\hat{x}(t)$ of the model:

(1) $0 < b_1 \leq \hat{x}_i(t) \leq b_2$ for constants b_1 and b_2;

(2)   each $e_{ij}(t)r_{ij}(t)$ should be of constant sign (always positive, always negative or always zero);

(3)   each food web has one-way energy flow (note that by condition (2) the predation communities do not change with time);

(4)   each $-a_i(t) \leq -\alpha < 0$;

(5)   if compartment j preys upon compartment i then the energy conversion efficiency $e_{ji}(t)/[-e_{ij}(t))]$ must be proportional at all time t to $\hat{x}_i(t)/\hat{x}_j(t)$;

(6)   each predation community has balanced predation loops;

(7)   each detritus gain term $d_i(x,t)$ must be uniformly continuous in time at $\hat{x}(t)$ (roughly speaking, d_i as a function cannot jump in value near $\hat{x}(t)$ or even change arbitrarily rapidly near $\hat{x}(t)$ as $t \rightarrow +\infty$, a quite reasonable assumption.

Attractor Trajectory Theorem. If a holistic ecosystem model as defined above fulfills conditions (1) through (7), then $\hat{x}(t)$ is an attractor trajectory for the model. The basin of attraction is the positive orthant and all of time.

Sketch of Proof (See the reference cited at the beginning of this section for details). Suppose first that the food web of the model consists of a single predation community. Conditions (2), (5), and (6) enable us to specify constants $\lambda_1, \lambda_2, ..., \lambda_n$ satisfying

$$\lambda_i e_{ij}(t)r_{ij}(t)\hat{x}_j(t) = -\lambda_j e_{ji}(t)r_{ji}(t)\hat{x}_i(t)$$

for all pairs of indices i,j (of course, this equation is trivial if $r_{ij}(t) = r_{ji}(t) = 0$, as is always the case of $i = j$). The constant λ_1 is defined to be 1. Such λ_i lead to the function of state space-time

$$\Lambda(x,t) = \sum_{i=1}^{n} \lambda_i \left[\frac{x_i}{\hat{x}_1(t)} - 1 - \ln\left(\frac{x_i}{\hat{x}_1(t)}\right) \right]$$

The rate of change of Λ along an arbitrary trajectory $x(t) \neq \hat{x}(t)$ can be shown by a lengthy calculation to satisfy

$$\dot{\Lambda}(x(t),t) < \sum_{i=1}^{n} -\alpha\lambda_i \frac{\left(x_i(t) - \hat{x}_i(t)\right)^2}{b_2}$$

Considering problems 3.13, 14, 15 and the bounds imposed by condition (1) on variations in components of $\hat{x}(t)$, conditions 1 and 2 of the Lyapunov Theorem can be shown to be fulfilled. If λ is the smallest number of the set of numbers $\lambda_1, \lambda_2, ..., \lambda_n$, then outside the level cylinder of radius ϵ about $\hat{x}(t)$ we have $\dot{\Lambda} \leq -n\alpha\lambda\epsilon^2/b_2$. Thus condition 3 of the Lyapunov Theorem is satisfied and $\hat{x}(t)$ is an attractor trajectory.

Now suppose that in the food web of the model there are two predation communities, labeled I and II, with n_1 and n_2 compartments. Suppose predation community I is upstream of predation community II. For predation community I let a Lyapunov function Λ_1 be defined as

$$\Lambda_1(x,t) = \sum_{i=1}^{n} \lambda_i \left[\frac{x_i}{\hat{x}_i(t)} - 1 - \ln\left(\frac{x_i}{\hat{x}_i(t)}\right) \right]$$

The previous argument can be used to prove that the first n_1 components of $\hat{x}(t)$ serve as an attractor trajectory for the subsystem corresponding to predation community I. It remains for us to consider the fate of the downstream subsystem.

Let a Lyapunov function for predation community II be defined by

$$\Lambda_2(x,t) = \sum_{i=n_1}^{n_1+n_2} \lambda_i \left[\frac{x_i}{\hat{x}_i(t)} - 1 - \ln\left(\frac{x_i}{\hat{x}_i(t)}\right) \right]$$

It can be shown that

$$\dot{\Lambda}_2 < \sum_{i=n_1}^{n_1+n_2} -\frac{\alpha\,\lambda_i\left(x_i - \hat{x}_i(t)\right)^2}{b_2} + \sum_{i=n_1}^{n_1+n_2} \frac{\lambda_i}{\hat{x}_i^2(t)x_i}\ (x_i-\hat{x}_i(t))\left[\hat{x}_i(t)\,d_i(x,t)-x_i d_i(\hat{x},t)\right]$$

The first sum is just what we had for the single predation community. Each term in the second sum is negative unless $x_i/\hat{x}_i(t)$ is closer to unity than $d_i(x,t)/d_i(\hat{x}(t),t)$.

Using condition (7) one can show this is impossible unless

$$\left(x_{n_1+1}(t), x_{n_1+2}(t),\dots,x_{n_1+n_2}(t)\right)$$

is within a certain collapsing zone around

$$\left(\hat{x}_{n_1+1}(t), \hat{x}_{n_1+2}(t),\dots,\hat{x}_{n_1+n_2}(t)\right)$$

For the downstream subsystem conditions 1 and 2 of the Lyapunov Theorem are fulfilled as before. Outside the collasping zone around $\hat{x}(t)$ condition 3 is fulfilled as well. Thus $\hat{x}(t)$ must be an attractor trajectory.

The above line of reasoning extends readily to any other downstream predation communities.

While the seven conditions are sufficient to guarantee $\hat{x}(t)$ is an attractor trajectory, they are by no means necessary. For example, condition (2) implies that the directed graph associated with each predation community must be constant with respect to time. This precludes migration, various forms of hibernation, and many other phenomena such as the seasonal emergence of insects. However, condition (2) can be modified since in proving that x(t) is an attractor trajectory we really only need n constants $\lambda_1, \lambda_2,\dots,\lambda_n$ such that

$$\lambda_i e_{ij}(t) r_{ij}(t)\hat{x}_j(t) = -\lambda_j e_{ji}(t) r_{ji}(t)\hat{x}_i(t)$$

This equation is, of course, trivially satisfied if $r_{ij}(t) = r_{ji}(t) = 0$. Thus models with predation interactions which turn on and off could actually be allowed with suitable constants λ_i.

Gain to a detritus compartment can occur as a by-product of predation, that is, as prey biomass not utilized by a predator. Suppose compartment 2 preys upon compartment 1 and suppose some energy resulting from predation flows into a detritus compartment, compartment 3. We would then have $r_{12}(t) = r_{21}(t) > 0$, $e_{12}(t) < 0$, and $e_{21}(t) > 0$. Also $d_3(x,t)$ would be positive and would depend upon the variables x_1 and x_2 (and possibly other state variables) as well as t.

Condition (3) precludes qualitatively getting something from nothing, an energy flow loop with a net gain. Condition(3) is a mathematical ideal; nature does recycle energy in limited quantities. For example, the lake ecosystem model in problem 5.14 recycles energy between nekton and a benthic detritus compartment. Here quantitative limitations on the magnitudes of "illegal" energy flow might save model stability.

REFERENCE

[J] C. Jeffries, Stability of holistic ecosystem models, in *Theoretical Systems Ecology*, ed. E. Halfon, Academic Press, New York, 1979, Chapter 20.

PROBLEMS

1. Suppose the dynamics of a simplified predator-prey model are given by

$$\dot{x}_1 = -x_1 x_2 - .1x_1^2 + 22x_1 - x_1 - x_1$$
$$\dot{x}_2 = .1x_1 x_2 - .1x_2^2 - 8x_2 - x_2$$

where $e_{12} = -1$; $r_{12} = r_{21} = 1$; $a_1 = .1$; $p_1 = 22$; $h_1 = 1$; $l_1 = 1$; $e_{21} = .1$; $a_2 = .1$; $h_2 = 8$; and $l_2 = 1$. Show $\hat{x}(t) = (100,10)$ is a constant trajectory for this system. Find λ_2 (using $\lambda_1 = 1$) so $\lambda_1 e_{12} r_{12} \hat{x}_2 = -\lambda_2 e_{21} r_{21} \hat{x}_1$. Write out the Lyapunov function used in the Attractor Trajectory Theorem as applied to this model.

2. Verify that the model in problem 1 fulfills the seven conditions of the Attractor Trajectory Theorem.

3. Suppose the model in problem 1 is modified as follows. Suppose $r_{12}(t) = r_{21}(t)$ is any continuous function of time with values greater than 5, $p_1(x,t) = 10r_{12}(t)+12$, and $h_2(t) = 10r_{12}(t)-2$. Is $\hat{x}(t) = (100,10)$ still an attractor trajectory?

4. Suppose the model of problem 3 is modified as follows. Suppose a_1 depends on x_1 in that: if $x_1 < 100$, then $a_1(x_1,t) = .01$; and if $x_1 \geq 100$, then $.01 \leq a_1(x_1,t) < .11$. Suppose h_1 becomes $h_1(x_1,t) = 11-100a_1(x_1,t)$. Is $\hat{x}(t) = (100,10)$ still an attractor trajectory?

5. Consider the two-dimensional predator-prey model

$$\dot{x}_1 = x_1 e_{12}(t) r_{12}(t) x_2 + p_1(t) x_1 - a_1(t) x_1^2 - h_1(t) x_1$$
$$\dot{x}_2 = x_2 e_{21}(t) r_{21}(t) x_1 - a_2(t) x_2^2 - h_2(t) x_2$$

with the following coefficient functions: $e_{12}(t) = -(2+\cos(t))^{-1}$; $e_{21}(t) = 3(2+\sin(t))^{-1}$; $h_1(t) = 1$; $r_{12}(t) = r_{21}(t) = 1$; $p_1 = 3+\cos(t)(2+\sin(t))^{-1}$. Show that $\hat{x}(t) = (2+\sin(t), 2+\cos(t))$ is a trajectory for this system.

6. Applying the Lyapunov function in the Attractor Trajectory Theorem to the model in problem 5, write out Λ and $\dot{\Lambda}$ explicitly as functions of

x_1, x_2, and t.

7. For the system in problem 6, use the Attractor Trajectory Theorem to prove $\hat{x}(t)$ is an attractory trajectory.

8. We now consider a hypothetical pelagic Antarctic ecosystem model somewhat analogous to the carbon flow model of Katherine A. Green (see "Systems approach to continental shelf ecosystems" by Bernard C. Patten and John T. Finn, in *Theoretical Systems Ecology*, ed. E. Halfon, Academic Press, New York, 1979). The model compartments are as follows.

1	ice algae
2	herbivorous ice invertebrates
2'	carnivorous ice invertebrates
3	phytoplankton
4	herbivorous zooplankton
4'	carnivorous zooplankton
5	fish
6	cephalopods
7	penguins
8	seals
9	whales

A realistic model would also feature several types of detritus compartments and, as devised by Green, a dissolved organic carbon compartment. For our present purpose, however, let us suppose that the main predation interactions are as follows: 1 is preyed upon by 2 and 4; 2 is preyed upon by 2', 4', and 5; 2' is preyed upon by 5; 3 is preyed upon by 4; 4 is preyed upon by 4', 5, 6, and 9; 4' is preyed upon by 5, 6, and 9; 5 is preyed upon by 6, 7, 8, and 9; 6 is preyed upon by 9; 7 is preyed upon by 8; and 8 is preyed upon by 9. Sketch the associated food web.

9. Suppose a holistic ecosystem model contains the predation interactions described in problem 8 in one predation community. Give an example of a relation among energy conversion efficiencies which would be consistent with balanced predation loops. If the compartments in a predation loop consist entirely of animal species and if the two chains comprising the loop are of unequal lengths (length

meaning number of compartments in a chain) then it seems unlikely that the loop will be balanced. An overly inclusive omnivore compartment would be especially likely to be part of such a loop. Can you suggest a stabilizing change involving the predation loop comprised of the 4,9 chain and the 4,5,9 chain?

ANSWERS TO PROBLEMS

1. Filling in $x_1 = 100$ and $x_2 = 10$ leads to

$$0 = -100 \cdot 100^2 + 22 \cdot 100 - 100 - 100$$
$$0 = .1 \cdot 100 \cdot 10 - .1 \cdot 10^2 - 8 \cdot 10 - 10$$

Hence $\hat{x}(t) = (100,10)$ is indeed a constant trajectory for the model. We need λ_2 so $1 \cdot (-1) \cdot 1 \cdot 10 = -\lambda_2 \cdot .1 \cdot 1 \cdot 100$, that is, $\lambda_2 = 1$. Thus

$$\Lambda = (x_1/100) - 1 - \ln(x_1/100) + (x_2/10) - 1 - \ln(x_2/10)$$

The model does fulfill all the conditions (with, say, $\alpha = .1$) of the Attractor Trajectory Theorem. Hence $\hat{x}(t)$ is an attractor trajectory with the entire positive orthant of (x_1, x_2)-space as basin of attraction. In particular,

$$\dot{\Lambda} < -.1(x_1 - 100)^2/100 - .1(x_2 - 10)^2/10 \text{ if } (x_1, x_2) \neq (100, 10).$$

2. Since $\hat{x}(t)$ is a constant trajectory, condition (1) is certainly fulfilled. Since e_{12}, e_{21}, and $r_{12} = r_{21}$ are constants, condition (2) is fulfilled. Since there is only one predation community in the model, condition (3) is fulfilled trivially. Since e_1 and e_2 are constants, condition (4) is fulfilled. Since both the energy conversion efficiency and the ratio $| x_2(t)/x_1(t) |$ are constants, condition (5) is fulfilled. Since there are no predation loops in the model, condition (6) is fulfilled trivially. Since there are no detritus gain terms in the model, condition (7) is fulfilled trivially.

3. Since

$$0 = -r_{12} \cdot 100 \cdot 10 - .1 \cdot 100^2 + (10r_{12} + 12) \cdot 100 - 100 - 100$$

and

$$0 = .1 \cdot r_{21} \cdot 10 \cdot 100 - .1 \cdot 10^2 - (10r_{21} - 2) \cdot 10 - 10$$

it is still true that $\hat{x}(t) = (100,10)$ is a constant trajectory for the system. All the conditions of the Attractor Trajectory Theorem remain fulfilled, so $\hat{x}(t)$ is an attractor trajectory.

142

4. The equation for $\dot{x}_1$ is now

$$\dot{x}_1 = -r_{12}x_1x_2 - a_1x_1^2 + (10r_{12}+12)x_1 - (11-100a_1)x_1 - x_1$$

Substituting $x_1 = 100$ and $x_2 = 10$ shows that $\hat{x}(t) = (100,10)$ is indeed a constant trajectory. Note $h_1 \geq 0$, as required for heat loss. Using $\lambda_1 = \lambda_2 = 1$ yields $\dot{\Lambda} \leq -1 \cdot .01 \cdot (x_1-100)^2/100 - 1 \cdot .01 \cdot (x_2-10)^2/10$ except at $\hat{x}(t)$ itself. Thus the Attractor Trajectory Theorem can still be used to establish that $\hat{x}(t)$ is an attractor trajectory.

5. We see that $\hat{x}(t) = (2+\sin(t), 2+\cos(t))$ is indeed a trajectory for the given system because substitution and some algebraic simplifications yield

$$\dot{x}_1 = \cos(t) = -(2+\sin(t))+3(2+\sin(t))+\cos(t)-2-\sin(t)-2-\sin(t)$$
$$\dot{x}_2 = -\sin(t) = 3(2+\cos(t))-(2+\cos(t))-2(2+\cos(t))-\sin(t)$$

6. To satisfy $\lambda_1 e_{12}r_{12}\hat{x}_2(t) = \lambda_2 e_{21}r_{21}\hat{x}_1(t)$ we require $\lambda_1 = 1$ and $\lambda_2 = 1/3$. Thus

$$\Lambda = \frac{x_1}{2+\sin(t)} -1-\ln\left(\frac{x_1}{2+\sin(t)}\right) + \frac{1}{3}\left[\frac{x_2}{2+\cos(t)} -1-\ln\left(\frac{x_2}{2+\cos(t)}\right)\right]$$

and

$$\dot{\Lambda} = -\frac{\left[x_1-\left(2+\sin(t)\right)\right]^2}{\left[2+\sin(t)\right]^2} - \frac{1}{3}\frac{\left[x_2-\left(2+\cos(t)\right)\right]^2}{\left[2+\cos(t)\right]^2}$$

7. Each component of $\hat{x}(t)$ has values between 1 and 3. Both $e_{12}r_{12}$ and $e_{21}r_{21}$ are of constant sign. There is only one predation community for this model. The auto-regulation functions satisfy $a_i \geq 1/3$. The energy conversion efficiency is 3 times the ratio $\hat{x}_2(t)/\hat{x}_1(t)$. There are no predation loops and no detritus compartments in this model. Thus the Attractor Trajectory Theorem can be used to prove $\hat{x}(t)$ is an attractor trajectory.

8. A sketch of the food web for the model in problem 8 is shown in Figure 6.1.

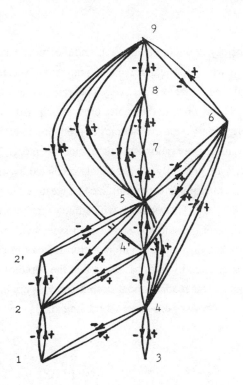

Figure 6.1. The food web of a hypothetical pelagic Antarctic ecosystem model.

9. In the model of problem 8, zooplankton are eaten by "whales" and zooplankton are eaten by fish, which are eaten by "whales." If all the associated predation efficiencies of this predation loop are roughly equal, then the predation loop will not be balanced. Thus for the sake of model stability it might be desirable to partition the "whales" compartment into three groups: whalebone whales (planktonophagous); fish hunters (ichthyophagous); and whales that feed largely on cephalopods (teuthophagous).

CHAPTER SEVEN
THE BEHAVIOR OF MODELS WITH ATTRACTOR REGIONS

7.1 Attractor Regions

Heretofore the object of our calculations and theorems has been the elucidation of models which for qualitative and quantitative reasons have attractor trajectories. To apply the Attractor Trajectory Theorem to the holistic ecosystem model one must somehow estimate the values of an anticipated attractor trajectory and ensure that energy conversion efficiencies of predation interactions are proportional to predator to prey ratios on the attractor trajectory. The fact that real ecosystem trajectories are generally not accurately predicted by such models is sometimes blamed on random perturbations such as weather factors.

There is another way. All our work with attractor trajectories has actually been only a foundation for the study of ecosystem models with attractor regions.

The distinction between models with attractor trajectories and models with attractor regions is suggested by Fig. 7.1. Note that the attractor region R is taken as independent of time. This mathematical simplification is not likely to significantly affect the theory of ecosystem modeling and is used in all that follows.

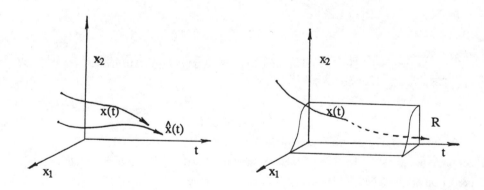

Figure 7.1. A typical trajectory x(t) approaches an attractor trajectory $\hat{x}$(t) and an attractor region R.

A set of points in state space is said to have diameter $\leq D$ if D is a number greater than or equal to the distance between any two points in the set; the *diameter of a set* is the smallest such D.

Suppose a set in state space is defined as the set of points satisfying an inequality $F(x) \leq 0$ where F is a multinomial in x, that is, a sum of products of powers of the components of x. Such a set or the intersection of a finite collection of such sets is called a *closed set*. In three-dimensional space the set of points satisfying $x_1^2 + x_2^2 + x_3^2 - 1 \leq 0$ is a closed set, the unit ball. Likewise in three-dimensional space the set of points satisfying $x_1 \geq 0$, $x_2 \geq 0$, $x_3 \geq 0$, and $x_1 + x_2 + x_3 - 1 \leq 0$ is a closed set, a pyramid in the nonnegative orthant with apex at the origin.

Suppose we are given a dynamical system and a closed set R of finite diameter in the nonnegative orthant of state space. Let R^+ denote the part of R in the positive orthant. In this book we call R an *ecosystem attractor region* with *basin of attraction S* provided:

1. if x(t) is any trajectory for the dynamical system starting at $x(t_0)$ in R^+, then x(t) must lie in R^+ for all finite $t > t_0$;

2. if x(t) is any trajectory starting at $x(t_0)$ in S but not in R, then there must exist a time interval T such that $t-t_0 > T$ implies x(t) is in R;

3. the time interval T in condition 3 must depend only on the distance from $x(t_0)$ to R, not $x(t_0)$ itself or t_0. (The distance from $x(t_0)$ to R is the greatest number δ such that the distance from $x(t_0)$ to any point in R is at least δ.)

If R is an attractor region with basin of attraction S and Q is a closed set in the nonnegative orthant with R contained in Q, then Q is also an attractor region with basin of attraction S.

Of course, the smaller the diameter of R is, the sharper the predictive capability of the model will be.

The object of this chapter and the next is to discuss the significance and machinery of attractor regions and then to describe qualitatively how to build ecosystem models with attractor regions. Generally the quantitative determination of the real usefulness of such models is a matter for computer simulation.

Consider the predator-prey type model

$$\dot{x}_1 = x_1 - x_1 x_2$$
$$\dot{x}_2 = x_1 x_2 - x_2^2$$

It is possible to prove that this model has a constant attractor trajectory, namely $\hat{x}(t) = (1,1)$. Any trajectory starting in the positive orthant at any time asymptotically approaches $\hat{x}(t)$. Also, any trajectory can be guaranteed to stay arbitrarily close to $\hat{x}(t)$ just by requiring that it start sufficiently close to $\hat{x}(t)$. Yet this model has no attractor region. If R is any closed set of finite diameter in the nonnegative orthant, let $x_1(t_0)$ be sufficiently large so that $x(t_0)$ is necessarily outside R. The trajectory $x(t)$ starting at $x(t_0)$ takes arbitrarily long to enter R as $x_2(t_0)$ varies from 1 to 0, even though the distance from R to any such initial point would be bounded. This is suggested by the typical trajectories shown in Fig. 7.2.

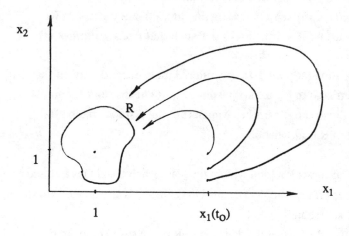

Figure 7.2. **Typical trajectories for $\dot{x}_1 = x_1 - x_1 x_2$, $\dot{x}_2 = x_1 x_2 - x_2^2$. The trajectories start at $(x_1(t_0), \varepsilon)$ with $0 < \varepsilon < 1$. As ε approaches 0, arbitrarily more time is needed to enter R. However, all such trajectories start within a finite distance of R. Therefore no set R can be an attractor region for the system even though the constant trajectory (1,1) is an attractor trajectory.**

Thus models with attractor trajectories may or may not have attractor regions. That models with attractor regions may or may not have attractor trajectories is shown in the next section. Attractor regions and attractor trajectories are associated with different types of stability.

7.2   The Lorenz Model

In 1963 Edward N. Lorenz published a revolutionary mathematical paper, "Deterministic Nonperiodic Flow" [L]. In it he reported on analytic and computer studies of a highly simplified, three-dimensional model of the dynamics of a fluid in a gravitational field between horizontal planes of different temperatures. The meteorological application of such a model would be the study of the onset of convective turbulence, as when the warmth of the earth's surface initiates large scale convection in the atmosphere.

As stated in problem 20 of Chapter 3, the model studied by Lorenz was

$$\dot{x}_1 = -10x_1 + 10x_2$$
$$\dot{x}_2 = -x_1x_3 + 28x_1 - x_2$$
$$\dot{x}_3 = x_1x_2 - (8/3)x_3$$

Let us consider, as did Lorenz, the function $\Lambda(x) = x_1^2 + x_2^2 + (x_3-38)^2 - 38^2$. The level sets of Λ are spheres in (x_1,x_2,x_3)-space centered at $(0,0,38)$. The rate of change of Λ along any trajectory of the system is

$$\dot{\Lambda} = 2x_1\dot{x}_1 + 2x_2\dot{x}_2 + 2(x_3-38)\dot{x}_3$$
$$= -20x_1^2 - 2x_2^2 - (16/3)(x_3-19)^2 + 5776/3$$

Thus outside the ellipsoid $20x_1^2 + 2x_2^2 + (16/3)(x_3-19)^2 = 5776/3$, $\dot{\Lambda}$ is negative. Outside the slightly larger ellipsoid E given by $20x_1^2 + 2x_2^2 + (16/3)(x_3-19)^2 = (5776/3)+1$, $\dot{\Lambda}$ is less than -1.

Let R be the set of points in (x_1,x_2,x_3)-space on or inside the (spherical) level set of Λ containing the point

$$(0,0,19-(5779/16)^{1/2})$$

(a point in E). One can use analytic geometry to show that R is a ball centered at

(0,0,38) with approximate radius 38.005. The region R is sketched in Fig. 7.3.

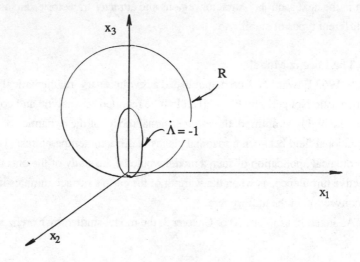

Figure 7.3. The relative positions of an attractor region R and the ellipsoidal set $\dot{\Lambda} = -1$ for the Lorenz system.

Now let us consider an arbitrary trajectory x(t) for the Lorenz system starting at some point $x(t_0)$ outside R. Let $\Lambda(x(t_0))$ be the value of Λ at $x(t_0)$. It happens that all other points the same distance as $x(t_0)$ from R have the same Λ value, a mathematically simplifying consequence of the choice of Λ. Let λ be the value of Λ on the surface of R. It follows that at most $[\Lambda(x(t_0))-\lambda]/1$ time units can elapse before x(t) or any other trajectory starting the same distance as $x(t_0)$ from R must enter R. The fact that $\dot{\Lambda} \leq -1$ on the surface of R means geometrically that every trajectory passing through R punctures R "inwardly." Thus R is an attractor region.

Once a trajectory enters R, what can happen? To find out, let us describe the constant trajectories of the Lorenz system. Of course, any constant trajectory must be inside R since $\dot{\Lambda} < -1$ outside R and since on a constant trajectory $\dot{\hat{x}}(t) = \hat{x}$ we have $\dot{\Lambda}(\hat{x}) = 0$.

A constant trajectory $\hat{x}$ satisfies

$$0 = -10\hat{x}_1 + 10\hat{x}_2$$

$$0 = -\hat{x}_1\hat{x}_3 + 28\hat{x}_1 - \hat{x}_2$$
$$0 = \hat{x}_1\hat{x}_2 - (8/3)\hat{x}_3$$

Since the first equation implies $\hat{x}_1 = \hat{x}_2$, we consider

$$0 = -\hat{x}_1\hat{x}_3 + 27\hat{x}_1 = \hat{x}_1(27 - \hat{x}_3)$$
$$0 = \hat{x}_1^2 - (8/3)\hat{x}_3$$

Clearly $(0,0,0)$ is a constant trajectory. If $\hat{x}_1 \neq 0$, then $\hat{x}_3 = 27$. If $\hat{x}_3 = 27$, then

$\hat{x}_1 = \hat{x}_2 = \pm\sqrt{72} = \pm 6\sqrt{2}$. Thus the constant trajectories of the Lorenz system are

$(0,0,0)$, $\left(6\sqrt{2}, 6\sqrt{2}, 27\right)$, and $\left(-6\sqrt{2}, -6\sqrt{2}, 27\right)$.

Let us determine the stability of the constant trajectories with the Hurwitz test (section five of Chapter Four). The linear approximation matrix of the Lorenz system is

$$A = \begin{pmatrix} -10 & 10 & 0 \\ -x_3 + 28 & -1 & -x_1 \\ x_2 & x_1 & -8/3 \end{pmatrix}$$

Evaluation at $(0,0,0)$ yields

$$A = \begin{pmatrix} -10 & 10 & 0 \\ 28 & -1 & 0 \\ 0 & 0 & -8/3 \end{pmatrix}$$

The characteristic polynomial here is $(z^2 + 11z - 270)(z + 8/3)$, the roots of which are $-8/3$ and (approximately) -22.83 and 11.83. The existence of a positive root implies by the Linear Approximation Theorem that $(0,0,0)$ is not an attractor trajectory for the full system.

Evaluation at $\left(6\sqrt{2}, 6\sqrt{2}, 27\right)$ yields

$$A = \begin{pmatrix} -10 & 10 & 0 \\ 1 & -1 & -6\sqrt{2} \\ 6\sqrt{2} & 6\sqrt{2} & -8/3 \end{pmatrix}$$

The characteristic polynomial of this matrix is $x^3+(41/3)x^2+(304/3)x +1440$. The second determinant of the Hurwitz test is

$$\det\begin{pmatrix} 41/3 & 1440 \\ 1 & 304/3 \end{pmatrix} \cong -55.11$$

Since a Hurwitz determinant is negative, $\left(6\sqrt{2},6\sqrt{2},27\right)$ cannot be an attractor trajectory.

A similar calculation shows that $\left(-6\sqrt{2},-6\sqrt{2},27\right)$ is not an attractor trajectory.

In summary, we know trajectories for the Lorenz system starting far from the origin must enter R. Inside R there are only three constant trajectories, none of which is an attractor trajectory. Computer simulations using the double-approximation procedure (section 2 of Chapter Three) show typical trajectories enter R well enough but thereafter seem to approach a sort of twisted surface, a "strange attractor." The surface is somewhat like a figure eight race track for toy cars; trajectories irregularly circle in one end of the track or cross over to the other end. Thus the strange attractor is, as a point set, quite different from a constant trajectory or a limit cycle. A schematic diagram of the Lorenz attractor is shown in Fig. 7.4.

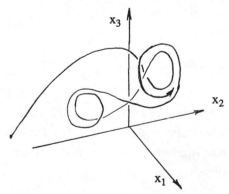

Figure 7.4. A typical trajectory approaches the figure eight strange attractor of the Lorenz system.

The Lorenz model is, relative to most other models of complex natural phenomena, extremely simple. It is now known that models with chaotic dynamics are not at all rare in three or more dimensions. In retrospect it is amazing the dynamical system experts were quite unaware of even the possibility of chaotic dynamics until about 1975. Much mathematical effort has subsequently been devoted to understanding chaos, but all the essential elements of the theory are represented in the original 1963 paper.

We can define chaotic dynamics as follows.

By *abstract attractor region* R (with basin of attraction S) we mean a closed set in the state space of an abstract dynamical system which meets conditions 2 and 3 of the ecosystem attractor region characterization (page 145). Suppose there is a positive number δ and that the trajectories of the system have a certain divergence property measured in terms of δ. We require that, given any positive number ε and any trajectory $x(t)$ starting at $x(t_0)$ in R, there exists a second trajectory $y(t)$ with $d(x(t_0),y(t_0)) < \varepsilon$ such that $d(x(t),y(t)) > \delta$ for an infinite sequence of positive time values $t_1 < t_2 < t_3 \ldots$. We require also that the time values are spaced as $t_{i+1} - t_i > \eta$, where η is a positive constant. In this book such a dynamical system is said to have *chaotic dynamics*.

Thus, roughly speaking, an abstract dynamical system has chaotic dynamics in an abstract attractor region R provided: with any given trajectory $x(t)$ starting at t_0 in R are associated other trajectories which start arbitrarily close to $x(t_0)$ but which drift apart from $x(t)$ to a distance exceeding δ; the trajectories which start near $x(t_0)$ might by chance from time to time be separated from $x(t)$ by a distance less than δ, but thereafter they drift apart again.

A good popular account of chaotic dynamics is the article "Chaos" by James Crutchfield, J. Doyne Farmer, Norman Packard, and Robert Shaw [CFPS]. For an indication of types of chaos which are associated with simple three-dimensional models, see "Continuous Chaos--Four Prototype Equations" by Otto Rössler [R].

If an ecosystem model has chaotic dynamics, then quantitative considerations become paramount. A sufficiently small attractor region can be, from the predictive viewpoint, almost as good as a constant attractor trajectory. Some ecosystem features associated with relatively small attractor regions are mentioned in Chapter Eight. On the other hand, if a well documented model turns out to have a large attractor region and a strange attractor therein, predicting the long term future of the ecosystem might be inherently impossible.

7.3 Elementary Ecosystem Models with Chaotic Dynamics

Suppose two competing autotrophs with population densities x_1 and x_2 are consumed by an herbivore with population density x_3 according to the equations

$$\dot{x}_1 = -.01x_1x_3+(1-.001x_2)x_1-.001x_1^2$$
$$\dot{x}_2 = -.001x_2x_3+(1-.0015x_1)x_2-.001x_2^2$$
$$\dot{x}_3 = .005x_3x_1+.0005x_3x_2-x_3$$

Here, in the notation of the holistic ecosystem model, $e_{13}r_{13} = -.01$, $p_1 = 1-.001x_2$, $a_1 = .001$, $e_{23}r_{23} = -.001$, $p_2 = 1-.0015x_1$, $a_2 = .001$, $e_{31}r_{31} = .005$, $e_{32}r_{32} = .0005$, and $l_3 = 1$. All other coefficients are assumed to vanish. This model was proposed by R. R. Vance [V] and further studied by Michael E. Gilpin [G].

The Trajectory Trapping Theorem can be directly applied to prove that no trajectory of the Vance model which starts in the positive orthant can reach the finite boundary of the positive orthant after a finite time interval.

Consider the function $\Lambda(x) = x_1+x_2+2x_3$. The rate of change of Λ along an arbitrary trajectory $x(t)$ of the Vance model is

$$\dot{\Lambda} = x_1-.001x_1^2-.001x_1x_2+x_2-.0015x_1x_2-.001x_2^2-2x_3$$
$$\leq x_1-.001x_1^2+x_2-.001x_2^2-2x_3$$
$$= -.001(x_1-500)^2-.001(x_2-500)^3-2x_3+500$$

Thus $\dot{\Lambda}$ is negative (in the positive orthant) outside the paraboloid given by $(x_1-500)^2+(x_2-500)^2+2000x_3 = 500,000$. Moreover, $\dot{\Lambda} < -1$ outside the slightly larger paraboloid P given by $(x_1-500)^2+(x_2-500)^2+2000x_3 = 501,000$. Let R be the closed region in the nonnegative orthant bounded by the boundary of the positive orthant and the plane given by $x_1+x_2+2x_3 = C$ (a constant) which intersects P at exactly one point, that is, the plane $\Lambda = C$ supported by P. This geometric arrangement is illustrated in Fig. 7.5.

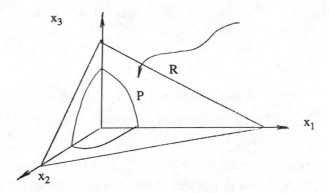

Figure 7.5. A typical trajectory approaches the attractor region R of the Vance system. The paraboloid P supports the upper surface of R.

Let $x(t)$ be any trajectory of the Vance system starting at $x(t_0)$ in the positive orthant. It follows that $x(t)$ and any other trajectory starting the same distance from R as $x(t_0)$ must enter R within $[\Lambda(x(t_0)) - C]/1$ time units. Since $\dot{\Lambda}$ is negative on the surface $\Lambda = C$ of R and since the Vance system fulfills the conditions of the Trajectory Trapping Theorem, no trajectory starting in R^+ can leave R^+ in finite time. Thus R is an attractor region for the Vance system.

There is exactly one constant trajectory $\hat{x}(t) = \hat{x}$ for the Vance system in the positive orthant. We require

$$0 = 1-.001\hat{x}_1 -.001\hat{x}_2 -.01\hat{x}_3$$
$$0 = 1-.0015\hat{x}_1 -.001\hat{x}_2 -.001\hat{x}_3$$
$$0 = -1+.005\hat{x}_1 +.0005\hat{x}_2$$

It follows that $\hat{x} = (4500/38,\ 2000-45000/38,\ [1000-\hat{x}_1-\hat{x}_2]/10\ )$
$\cong (118.42,\ 815.79,\ 6.5789)$. The linear approximation matrix for the Vance

system is

$$A = \begin{pmatrix} 1-.002x_1-.001x_2-.01x_3 & -.001x_1 & -.01x_1 \\ -.0015x_2 & 1-.0015x_1-.002x_2-.001x_3 & -.0001x_2 \\ .005x_3 & .0005x_3 & -1+.005x_1+.0005x_2 \end{pmatrix}$$

The characteristic polynomial of A evaluated at $\hat{x}$ is approximately $z^3+.934z^2$ $-.00667z+.02415$. Since the coefficient of z is negative, $\hat{x}$ cannot be an attractor trajectory.

The constant trajectories on the boundary of the positive orthant are (0,0,0), (1000,0,0), (200,0,80), and (0,1000,0). Of these, the first three are not attractor trajectories by virtue of the Hurwitz test. The Hurwitz test applied to (0,1000,0) is inconclusive. However, nonlinear methods can be used to show that (0,100,0) is not an attractor trajectory.

Thus the Vance model has an attractor region in the positive orthant but no attractor trajectories in the attractor region. Computer simulations performed by Gilpin indicate that the Vance model also has chaotic dynamics in common with the Lorenz model. A schematic diagram of the Vance attractor is shown in Fig. 7.6.

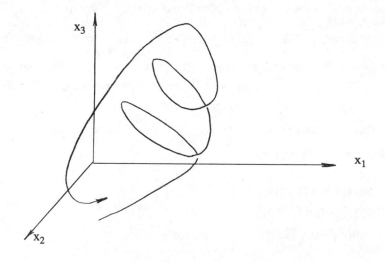

Figure 7.6. A typical trajectory near the strange attractor of the Vance system.

REFERENCES

[CFPS] J. Crutchfield, D. Farmer, N. Packard, and R. Shaw, Chaos, Scientific American 255 (December 1986) 46-57.

[G] M. E. Gilpin, Spiral chaos in a predator-prey model, American Naturalist 113 (1979) 306-308.

[L] E. Lorenz, Deterministic nonperiodic flow, Journal of the Atmospheric Sciences 20 (1963) 130-141.

[R] Otto Rössler, Continuous Chaos--Four Prototype Equations, Annals of the New York Academy of Sciences 316 (1979) 377-392.

[V] R. Vance, Predation and resource partitioning in one predator-two prey model community, American Naturalist 112 (1978) 797-813.

PROBLEMS

1. Consider the one-dimensional population model

$$\dot{x} = p(x)x - a(x)x^2$$

where the photosynthesis function $p(x)$ and the auto-regulation function $a(x)$ are continuous with $0 \le p(x) \le 1$ and $a(x) \ge 1$ for all $x \ge 1$. For $x < 1$, $p(x)$ and $a(x)$ are nonnegative but otherwise arbitrary. Show that any trajectory for this system which starts in the positive orthant $(x > 0)$ cannot reach the finite boundary of the positive orthant $(x = 0)$ in finite time. Using the inequality $\dot{x} \le -(x-1)$ for $x \ge 1$ show that the closed interval $R = [0,2]$ is an attractor region for this model.

2. Find a smaller attractor region for the model in problem 1.

3. Comment on the one-dimensional population model

$$\dot{x} = \frac{2(2+x)}{3(1+x)} x - [x^2 - x+1]x^2 = p(x)x - a(x)x^2$$

4. Consider the predator-prey model

$$\dot{x}_1 = -x_1 x_2 + p(x,t)x_1 - a_1(x,t)x_1^2$$

$$\dot{x}_2 = .1x_1 x_2 - a_2(x,t)x_2^2$$

where $0 \le p(x,t) \le 10$ and $1 \le a_i(x,t)$. Show the rate of change of $\Lambda(x) = x_1 + 10x_2$ along an arbitrary trajectory in the positive orthant is ≤ -1 outside the closed region R bounded by the boundary of the positive orthant and the level set $\Lambda = 5 + 2(2600/101)^{1/2}$. Show R is an attractor region with basin of attraction the entire positive orthant.

5. Consider the model of two competing autotrophs given by

$$\dot{x}_1 = \frac{2}{x_2 + 1} x_1 - a_1(x,t)x_1^2$$

where $1 \le a_i(x,t)$. Let $\Lambda(x) = (x_1+1)(x_2+1)$. Construct a closed set R in the nonnegative orthant outside which the rate of change of Λ along an arbitrary trajectory is ≤ 1. Show that R is an attractor region for the system and that the entire positive orthant is a basin of attraction for R.

6. Consider the two-dimensional dynamical system

$$\dot{x}_1 = -x_1 x_2$$

$$\dot{x}_2 = x_2^2$$

Does this model fulfill the conditions of the Trajectory Trapping Theorem? Verify that all trajectories for this model lie on arcs of the hyperbolae $x_1 x_2 = $ constant. Verify that a trajectory for the model starting at $t = 0$ and $x = (1,1)$ is $x(t) = (1-t,(1-t)^{-1})$. This trajectory has $x_1(1) = 0$. Why does not this value for $x_1(1)$ contradict the Trajectory Trapping Theorem?

7. Write a program for calculating trajctories of the Lorenz system using the double-approximation procedure.

8. Write a program for calculating trajectories of the Vance system using the double-approximation procedure.

ANSWERS TO PROBLEMS

1.   The trajectory Trapping Theorem implies that no trajectory for the system starting in the positive orthant can reach $x = 0$ in finite time. For $x \geq 1$, $\dot{x} = p(x)x - a(x)x^2 \leq x - x^2 = x(1-x) \leq -(x-1)$. Thus in the positive orthant but outside $R = [0, 2]$, $\dot{x} \leq -1$. A trajectory starting at $x(t_0)$ can take at most $[x(t_0)-2]/1$ time units to enter R. Since $\dot{x}(2) \leq -1$ (that is, any trajectory is decreasing in x value at $x = 2$), no trajectory starting in $R^+ = (0,2)$ can leave R^+. Thus R is an attractor region.

2.   Let $R = [0, 1+ \varepsilon]$ where ε is a number between 0 and 1. It follows that $\dot{x} \leq -\varepsilon$ for x in the positive orthant but outside R. Thus at most $[x(t_0)-(1+\varepsilon)]/\varepsilon$ time units can elapse before a trajectory starting at $x(t_0) > 1+\varepsilon$ must enter R. Since $\dot{x}(1+\varepsilon) \leq -\varepsilon$, no trajectory starting in R^+ can cross $x = 1+\varepsilon$. Thus R is an attractor region.

3. We note for $x \geq 1$ that $2/3 \leq p(x) \leq 1$. Also, for $x \geq 1$, we have $a(x) \geq 1$. Thus the answers for problems 1 and 2 can be applied to show $[0, 1+\varepsilon]$ (where ε is any positive number) is an attractor region for the model.

4. The rate of change of Λ along an artibrary trajectory is

$$\dot{\Lambda} = -x_1x_2 + p(t)x_1 - a_1(t)x_1^2 + x_1x_2 - 10a_2(t)x_2^2$$
$$\leq 10_1 - x_1^2 - x_2^2$$
$$= -(x_1-5)^2 - x_2^2 + 25$$

Thus $\Lambda \leq -1$ outside the positive orthant part of the circle $(x_1 -5)^2 + x_2^2 = 26$. This circle supports the line $x_1 + 10x_2 = 5 + 2(2600/101)^{1/2}$, as shown in Fig. 7.7. The Trajectory Trapping Theorem can be used to show that no trajectory starting in the positive orthant can leave the positive orthant in finite time. A trajectory starting at $x(t_0)$ outside R or at any other point not further from R than $x(t_0)$ must enter R after at most $[\Lambda(x(t_0))-(5+2(2600/101)^{1/2})]/1$ time units. Since the rate of change of Λ is less than or equal to -1 on the level set where $\Lambda = 5+2(2600/101)^{1/2}$, no trajectory starting in the positive orthant portion of R can cross the boundary of R. Thus R is an attractor region.

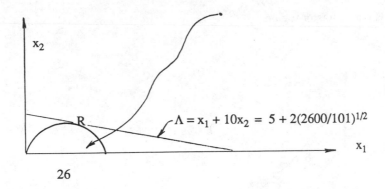

Figure 7.7. An attractor region for $\dot{x}_1 = -x_1x_2+p(x,t)x_1-a_1(x,t)x_1^2$; $\dot{x}_2 = .1\ x_1x_2-a_2(x,t)x_2^2$.

5. The Trajectory Trapping Theorem ensures that any trajectory for the system which starts in the positive orthant cannot reach the boundary of the positive orthant in finite time.

Along an arbitrary trajectory of the system the rate of change of Λ is

$$\dot{\Lambda} = \dot{x}_1(x_2+1)+\dot{x}_2(x_1+1)$$
$$\leq 2x_1-x_1^2+2x_2-x_2^2$$
$$= -(x_1-1)^2-(x_2-1)^2+2$$

Thus $\dot{\Lambda} \leq -1$ outside positive orthant portion of the circle $(x_1-1)^2+(x_2-1)^2 = 3$.

Let R be the closed set in the nonnegative orthant bounded by the boundary of the positive orthant and the hyperbolic level set of Λ supported by the circle $(x_1-1)^2 + (x_2-1)^2 = 3$, that is, the hyperbola $\Lambda = [2+(3/2)^{1/2}]^2$. Consider an arbitrary trajectory x(t) starting in the positive orthant at x(t$_0$) but outside R at a distance δ from R. Let Q be the set of points in the positive orthant at a distance δ from R, as sketched in Fig. 7.8. Let λ be the maximum value of Λ on Q. It follows that at most $[\lambda-(2+(3/2)^{1/2})^2]/1$ time units can elapse before x(t) or any other trajectory starting in the positive orthant as close to R as x(t$_0$) must enter R. Likewise no trajectory starting in R$^+$ can leave R$^+$ in finite time.

Thus R is an attractor region.

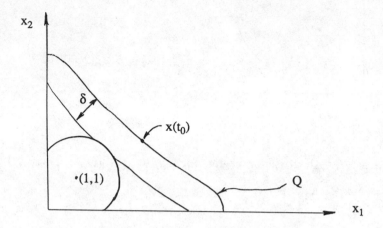

Figure 7.8. An attractor region for a model of two competing autotrophs.

6. The conditions of the Trajectory Trapping Theorem are fulfilled by the model. Using the notation in the statement of the theorem in section two of Chapter Three, $g_1(x,t) = -x_2$; $h_1(x,t) = 0$; $g_2(x,t) = x_2$; and $h_2(x,t) = 0$. For $0 < t < 1$, the given trajectory has

$$\dot{x}_1 = -1 = -(1-t)(1-t)^{-1} = -x_1 x_2$$
$$\dot{x}_2 = +(1-t)^{-2} = x_2^2$$

as required. Indeed, $\lim_{t \to 1} x_1(t) = 0$, but $\lim_{t \to 1} x_2(t) = +\infty$. The point $(0,+\infty)$ is not on the <u>finite</u> boundary of the positive orthant, so the trajectory does not contradict the theorem.

7. A Lotus 1-2-3 program for calculating trajectories of the Lorenz system with the double-approximation procedure is listed below. Note that recalculation is rowwise and the macro range is A1-A4. To start the calculation enter initial values for x_1, x_2, x_3 in A11, A12, A13. Then copy B11.B13 into A11.A13. The is run initiated with CTRL+Shift+A.

```
A1:  '{goto}B8~+B6+B8~
A2:  '{calc}
A3:  '/xi(B7-B8)>0~/xgA2~
A4:  '/xq
A5:  'delta t
A6:  0.01
A7:  'max t
B7:  10
A8:  'current t
B8:  +B6+B8
A11: 0
B11: +B21
A12: 0
B12: +B22
A13: 0
B13: +B23
A15: +A11+B6*(-10*A11+10*A12)
A16: +A12+B6*(-A11*A13+28*A11-A12)
A17: +A13+B6*(A11*A12-(8/3)*A13)
A21: +A11+0.5*(A15-A11)+0.5*B6*(-10*A15+10*A16)
A22: +A12+0.5*(A16-A12)+0.5*B6*(-A15*A17+28*A15-A16)
A23: +A13+0.5*(A17-A13)+0.5*B6*(A15*A16-(8/3)*A17)
```

8. A Lotus 1-2-3 program for calculating trajectories of the Vance system with
the double-approximation procedure is listed below. Note that recalculation is
rowwise and the macro range is A1--A4.

```
A1:  '{goto}B8~+B6+B8~
A2:  '{calc}
A3:  '/xi(B7-B8)>0~/xgA2~
A4:  '/xq
A5:  'delta t
A6:  0.01
A7:  'max t
B7:  10
A8:  'current t
B8:  +B6+B8
A11: +A21
B11: +B21
A12: +A22
B12: +B22
A13: +A23
B13: +B23
A15: +A11+B6*A11*(1-0.001*A11-0.001*A12-0.01*A13)
A16: +A12+B6*A12*(1-0.0015*A11-0.001*A12-0.001*A13)
A17: +A13+B6*A13*(-1+0.005*A11+0.0005*A12)
A21: +A11+0.5*(A15-A11)
     +0.5*B6*A15*(1-0.001*A15-0.001*A16-0.01*A17)
A22: +A12+0.5*(A16-A12)
     +0.5*B6*A16*(1-0.0015*A15-0.001*A16-0.001*A17)
A23: +A13+0.5*(A17-A13)+0.5*B6*A17*(-1+0.005*A15+0.005*A16)
```

CHAPTER EIGHT

HOLISTIC ECOSYSTEM MODELS WITH ATTRACTOR REGIONS

8.1 An Attractor Region Theorem

This section contains the attractor region analogue of the attractor trajectory result in section three of Chapter Six. Our aim is the specification of conditions on the coefficient functions of a very general type of ecosystem model which are sufficient to guarantee the existence of an attractor region.

Let us consider the holistic ecosystem model

$$\dot{x}_i = \sum_{j=1}^{n} x_i e_{ij}(x) r_{ij}(x,t) x_j - a_i(x,t) x_i^2 + p_i(x,t) x_i - h_i(x,t) x_i + d_i(x,t) - l_i(x,t) x_i$$

where

$r_{ij}(x,t) x_i x_j$, $i \neq j$, is the rate of predation interactions between compartments i and j;

$e_{ij}(x)$, with $i \neq j$, is the rate of gain ($e_{ij} > 0$) or loss ($e_{ij} < 0$) in compartment i per predation interaction with compartment j;

$-a_i(x,t) x_i^2$ for nonautotroph compartments is the rate of energy loss from compartment i due to intracompartmental predation, territorial conflicts, autolysis, or related mechanisms;

$p_i(x,t) x_i - a_i(x,t) x_i^2$ for autotroph compartments is the rate of potential photosynthetic gain minus intracompartmental competition loss, that is, net photosynthetic gain;

$-h_i(x,t) x_i$ is the rate of energy loss from compartment i to sinks outside the model including energy losses as heat, methane, or peat;

$d_i(x,t)$ for a detritus compartment is the rate of detritus gain from upstream compartments;

$-l_i(x,t) x_i$ is the rate of energy loss from compartment i to detritus compartments due to excretion, physical breakdown, or other nonpredation mechanisms.

As in Chapter Six, the coefficient functions r, e, a, p, h, d, and l are all assumed to be defined and to have continuous first derivatives for all time and throughout the nonnegative orthant. The Trajectory Existence Theorem then guarantees the existence of trajectories starting in the positive orthant at any time.

Considering the role of each type of coefficient function, we assume further that each r, e, a, p, h, d, and l is nonnegative. This assumption and the Trajectory Trapping Theorem automatically imply that any trajectory starting in the positive orthant cannot reach the finite boundary of the positive orthant in finite time.

Note that each e is generally dependent upon x, not t. Also each r, a, p, h, d, and l depends upon both x and t. In these ways the attractor region model is significantly more general than the attractor trajectory model.

Let us recall and slightly modify some concepts from the attractor trajectory theory. The food web of a model at fixed x and t is the signed digraph with nodes corresponding to system compartments and edges corresponding to predation interactions ("+,-" 2-cycles) and detritus donation (edges between predation communities). Specifically, if $e_{ij}(x,t) \neq 0$, then an edge occurs from node j to node i, signed as e_{ij}; if $\partial d_i/\partial x_j \neq 0$, then an edge occurs from node j to node i, signed as $\partial d_i/\partial x_j$. We assume throughout that if $e_{ij}(x,t) \neq 0$, then $\partial d_i/\partial x_j = 0$.

A *predation community* in a food web is a maximal connected subgraph the nodes of which are connected by "+,-" 2-cycles (predation links) only.

Suppose for some food web and some x and t that no p-cycle, $p \geq 3$, involves nodes in more than one predation community. In a natural way such a restriction implies a hierarchy of predation communities, a transitive partial ordering according to the pathways of energy flow. To be precise, if there exists a directed path from a node in predation community I to a node in predation community II (possibly involving nodes in intermediate predation communities), then predation community I is said to be *upstream* of predation community II, or, equivalently, predation community II is said to be *downstream* of predation community I. Suppose the predation communities of a system are independent of x and t, although the edges between predation communities might change with various x and t. Suppose the hierarchy is consistent for all x and t in the sense that if predation community I is upstream of predation community II for one x and t, then predation community I is not downstream of predation community II for any other x and t. Such a holistic model is said to have *one-way energy flow*.

If the net energy conversion efficiencies of different chains of predation interactions between two compartments are equal for all x, then the predation community is said to have *balanced predation loops*.

With all these definitions in mind, we are in a position to describe some
sufficient conditions for the existence of an attractor region.

**Attractor Region Theorem. The following conditions are sufficient
to imply the existence of an attractor region:**

(1) any $r_{ij}(x,t)e_{ij}(x)$ which is not identically zero must satisfy
$0 < b_1 \leq |e_{ij}(x)| \leq b_2$ for constants b_1 and b_2 and have $r_{ij}(x,t)$
of constant sign; if a system satisfies this condition, the food
web of the system is independent x and t;

(2) the food web of the system must have one-way energy flow;

(3) each predation community must have balanced predation loops
for all x and t;

(4) for each compartment i a constant $\varepsilon_i > 0$ must exist with either
$h_i(x,t) + l_i(x,t) \geq \varepsilon_i$ for all x and t, or $a_i(x,t) \geq \varepsilon_i$ for all x and
t;

(5) for each autotroph compartment i there must exist a positive
number P_i such that $p_i(x,t) \leq P_i$ for all x and t; furthermore, a
positive number β_i must exist such that $p_i(x,t) \leq \beta_i a_i(x,t)$
for all x and t;

(6) for each detritus compartment i and each x there must exist a
positive number $D_i(x)$ such that $d_i(x,t) \leq D_i(x)$ for all t.

Proof. We shall outline a procedure for constructing an attractor region for a
holistic model satisfying conditions (1) through (6). A detailed account appears in
[J].

First let us consider the case in which the system consists of a single
predation community with n compartments, at least one of which is an autotroph
compartment and none of which is a detritus compartment. We define n functions
$N_i(x)$ (that is, a vector field $N(x)$) as follows: let $N_1(x) = 1$; if node 1 is connected
to i (by a "+,-" 2-cycle) in the food web of the system, let $N_i(x)$ solve $N_i(x)e_{i1}(x) =$
$-N_1(x)e_{1i}(x)$; let the definition of N propagate through the food web so that for all i
$\neq j$

$$N_i(x)e_{ij}(x) = -N_j(x)e_{ji}(x)$$

This procedure consistently defines all components of N because of the predation loops condition, condition 3.

Consider the hypersurface H defined in the positive orthant to be the union of the following curves. Select n-1 numbers $K_2, K_3, ..., K_n$ with $K_2^2 + K_3^2 + ... + K_n^2 = 1$. For notational purposes let us regard such numbers as a point K in an abstract (n-1)-dimensional space. Let $\tau \geq 0$ be an independent variable parameterizing the curve

$$C_K(\tau) = (c_K(\tau), K_2\tau, K_3\tau, ..., K_n\tau) + x_0$$

where x_0 is a point on the ray starting at the origin 0 and passing through (1,1,...,1) and where $c_K(\tau)$ is a function with $c_K(0) = 0$ and

$$dc_K/d\tau = -K_2 N_2(C_K(\tau)) - K_3 N_3(C_K(\tau)) - ... - K_n N_n(C_K(\tau))$$

The existence and uniqueness theorem for first order differential equations guarantees that $c_K(\tau)$ exists and is unique (see [S, p. 8]). Uniqueness implies that no two such curves have any point in common except x_0.

The bounds on the functions $e_{ij}(x)$ in condition (1) imply the existence of bounds on each $N_i(x)$. Such bounds on $N(x)$ in turn imply the portion of each curve $C_K(\tau)$ in the nonnegative orthant is finite; in fact, the hypersurface H generated by fixing x_0 and allowing K to vary over all possible values has finite diameter. A typical curve and the hypersurface H of all such curves starting at a fixed x_0 is shown in Fig. 8.1. No two such hypersurfaces have any points in common.

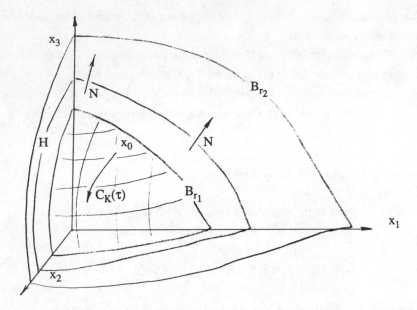

Figure 8.1 A construction used in the proof of the Attractor Region Theorem. For x_0 sufficiently far from the origin, the hypersurface H generated by curves $C_K(\tau)$ is always punctured inwardly by trajectories of the model.

The tangent vector of the curve $C_K(\tau)$ is

$$(dc_K/d\tau, K_2, K_3, \ldots, K_n) = (-K_2N_2 - K_3N_3 - \ldots - K_nN_n, K_2, K_3, \ldots, K_n)$$

This tangent vector is perpendicular to N. Thus, as shown in the figure, N is the outward pointing normal vector field of the surface generated by the curves.

For any positive number r let B_r be the set of points in the nonnegative orthant at a distance r from the origin 0. For any hypersurface H generated by curves like $C_K(\tau)$, there exist positive numbers r_1 and r_2 such that any x in H is between B_{r_1} and B_{r_2}, that is, $r_1 < d(x,0) < r_2$. Such B_{r_1} and B_{r_2} are shown in Fig. 8.1. Likewise, for any B_r there exist hypersurfaces H_1 and H_2 generated by curve families like $C_K(\tau)$ such that B_r lies between H_1 and H_2, that is, any ray in the nonnegative orthant starting at 0 passes first through H_1, then B_r, then H_2.

Let x(t) be an arbitrary trajectory for the model starting in the positive orthant. The dot product of $\dot{x}(t)$ and $N(x(t))$ is

$$\dot{x} \cdot N = \sum_{i,j=1}^{n} N_i e_{ij} r_{ij} x_i x_j + \sum_{i=1}^{n} N_i(-a_i x_i^2 + p_i x_i - h_i x_i - l_i x_i)$$

The first sum on the right is zero by virtue of our choice of components of N. By virtue of the bounds in conditions (4) and (5), we can select a positive number ε so that the second sum is $< -\varepsilon$ outside a region of finite diameter Q in the nonnegative orthant. Thus for x_0 sufficiently large, the trajectories of the system puncture the corresponding hypersurface H "inwardly" relative to 0. If a trajectory starts in a region bounded by such a hypersurface and the boundary of the positive orthant, then the trajectory can never leave that region. Thus every trajectory of the system is bounded in the future.

Let R be the closed set in the nonnegative orthant bounded by the boundary of the positive orthant and the unique hypersurface H generated by curves like $C_K(\tau)$ and supported by Q. Using a line of argument parallel to that in the proof of the Lyapunov Theorem of section three of Chapter Six, it follows that R must be entered by any trajectory starting in the positive orthant within a finite time interval T. Moreover, T does not depend upon the exact value of the initial state or the initial time; T depends only on initial distance to R. For any such distance there will exist a worst case, an initial state and time which results in maximum time T to enter R.

Next consider the case that the model is a single detritus based predation community. Thus in view of condition 2 (there is no upstream compartment in the model), each $d(x,t) = d(t)$. In view of condition 6, $0 \le d(t) \le D$. The procedure used in the previous case again leads to an attractor region.

For the general case of many predation communities, we need only consider predation communities as subsystems and one by one apply the above results to build an attractor region for each subsystem.

8.2 An Example

Let's conclude our treatment of attractor region stability with a simple but illustrative example. Suppose compartment 2 "preys upon" autotroph compartment 1 and that the by-product of this "predation" is donated to detritus compartment 3. Thus we have a model like the following

$$\dot{x}_1 = -r_{12}x_1x_2 - a_1x_1^2 + p_1x_1 - l_1x_1 - h_1x_1$$
$$\dot{x}_2 = 0.5r_{21}x_1x_2 - h_2x_2$$
$$\dot{x}_3 = -a_3x_3^2 + (0.5r_{12}x_1x_2)$$

Here $e_{12} = -1$ and $e_{21} = .5$, so the functions $N_1(x)$ and $N_2(x)$ are constants, namely, $N_1 = 1$ and $N_2 = 2$. Half the energy lost from compartment 1 per predation incident goes to compartment 2, half to compartment 3. We can suppose the autotrophs in compartment 1 are the "living" portions of a plant, the portions which contribute to or engage in photosynthesis and are eaten by herbivores from compartment 2. Thus $-l_1x_1$ might be energy contributed to "nonliving" portions of the same plant, a woody bole, for instance. The term $-h_1x_1$ refers to heat or respiration losses. Note that d_3 is $0.5r_{12}x_1x_2$, detritus gain due to predation between compartments 1 and 2.

Suppose the functions $r_{12} = r_{21}$, a_1, p_1, l_1, h_1, h_2, and a_3 are all continuous functions of x and t as shown in Fig. 8.2.

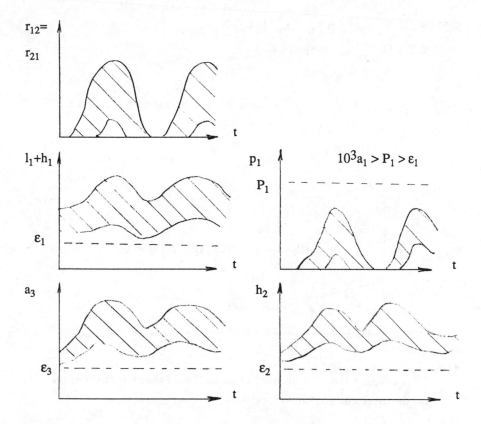

Figure 8.2. Admissible functions for the three compartment model.

Let's check that this model satisfies the conditions of the Attractor Region Theorem. Condition 1 is satisfied since e_{12} and e_{21} are constants. Condition 2 is satisfied because the predation community consisting of compartments 1 and 2 donates energy to compartment 3 (a predation community with one compartment). Condition 3 is trivially fulfilled since the predation communities have only one or two compartments. Condition 4 is fulfilled since the functions $l_1 + h_1$, h_2, and a_3 are as given in Fig. 8.2. Condition 5 is likewise fulfilled since p_1 is as shown in Fig. 8.2. The graph of $r_{12} = r_{21}$ in Fig. 8.2 is included to show $d_3 < R_{12} x_1 x_2$ so the relevant parts of conditions 2 and 6 are fulfilled. Therefore the above three-compartment model has an attractor region.

The size of the attractor region may be estimated as follows. As noted previously, $N_1 = 1$ and $N_2 = 2$. Also

$$\dot{x}_1 N_1 + \dot{x}_2 N_2 = -a_1 x_1^2 + p_1 x_1 - l_1 x_1 - h_1 x_1 - 2h_2 x_2$$
$$\leq P_1 x_1 (1 - 10^{-3} x_1) - \varepsilon_1 x_1 - 2\varepsilon_2 x_2$$

Thus $\dot{x}_1 N_1 + \dot{x}_2 N_2 \leq -\varepsilon_1 x_1 - \varepsilon_2 x_2$ "outside" the parabola shown in Fig. 8.3.

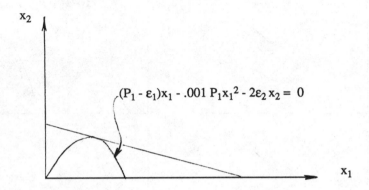

Figure 8.3. A two-dimensional cross section of the attractor region of the three compartment model.

Regardless of the behavior of x_3, variables x_1 and x_2 must enter the cross-hatched region of Fig. 8.2. Let D be (finite) maximum value $0.5 r_{12} x_1 x_2$ can have in that cross-hatched region. The setting $N_3 = 1$ (because compartment 3 starts another predation community) we have

$$\dot{x}_3 N_3 = -a_3 x_3^2 + 0.5 r_{12} x_1 x_2$$

Thus after a finite amount of time T determined by initial conditions, any trajectory satisfies

$$\dot{x}_3 N_3 \leq -\varepsilon_3 x_3^2 + D.$$

Let $\Lambda = x_1 + 2x_2 + x_3$. Thus after T time units,

$$\dot{\Lambda} = \dot{x}_1 N_1 + \dot{x}_2 N_2 + \dot{x}_3 N_3 \leq (P_1 - \varepsilon_1)x_1 - 10^{-3}P_1 x_1^2 - 2\varepsilon_2 x_2 - \varepsilon_3 x_3^2 + D$$

Consider the level set L of Λ in the nonnegative orthant supported by the paraboloid $(P_1 - \varepsilon_1)x_1 - 10^{-3}P_1 x_1^2 - 2\varepsilon_2 x_2 - \varepsilon_3^2 + D + 1 = 0$. Let R be the region bounded by L and the boundary of the positive orthant (Fig. 8.4). Thus in the positive orthant but outside R, $\dot{\Lambda} < -1$ after T time units. It follows that R is an attractor region for the system.

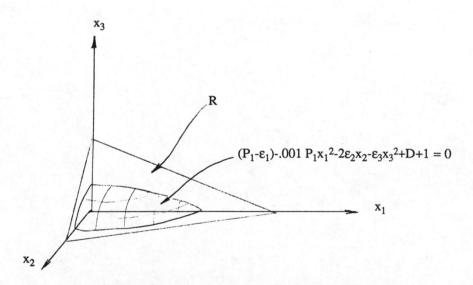

Figure 8.4. Explicit representation of the attractor region R for the three compartment model.

The Attractor Region Theorem is perhaps significant in that it provides a recipe for building stable models of large dimension. Actual calculation of attractor region size is generally not practical. Numerical simulations might be used to investigate the stability of a large model.

REFERENCES

[J] C. Jeffries, Ecosystem modelling: qualitative stability, *Systems and Control Encyclopedia*, ed. M. Singh, Pergamon, Oxford (1987) 1348-1351.

[S] David Sánchez, *Ordinary Differential Equations and Stability Theory: An Introduction,* Freeman, San Francisco, 1968.

CHAPTER NINE
SEQUENCING ENERGY FLOW MODELS TO ACCOUNT FOR SHORTGRASS PRAIRIE ENERGY DYNAMICS

9.1 Energy Flow and Accumulation Modeling

The first eight Chapters of this book have been devoted to an informal presentation of the mathematical foundations of ecosystem modeling theory. Some of the text and problems have been couched in terms of ecosystems, but we have dealt mainly with abstract mathematics. Even with "ecosystem" equations we have used energy flow rate functions (the "collisions" type predation rates) which would be better suited to idealized chemical reaction networks than to ecosystem modeling. Such is the nature of instruction.

The time has come to deliver a realistic model of actual ecosystem energy flows.

Spring comes to the upland prairie of southwestern Saskatchewan about 15 May of each year. The snow melts, the ground thaws. Grasses produce new shoots from roots, and insect eggs hatch.

Thousands of birds migrate north over the prairie, most en route to the boreal forest, the tundra, or the arctic coast. Among those which stay to attempt to utilize the energy resources of the prairie are the clay-colored sparrows, *Spizella pallida* (called hereafter "sparrows"). These sparrows nest out on open, rocky terrain and rely on camouflage to protect themselves from mammalian and avian predators. Hence they are exceedingly drab little birds. From the modeling viewpoint the sparrows have an outstanding feature: in areas with abundant grasshoppers, they are thought to eat little else during the period 15 May --15 July, the 61 days (1464 hours) of the sparrow reproductive season. What follows is an energy accounting model of the autotroph-grasshopper-sparrow system during those hours.

Our model will assume that aggregated energy is a biological currency. Nutritional and biochemical details of how food energy is utilized in an organism will be neglected. The energy in a compartment will be taken as the heat released when dried (12% moisture) biomass from that compartment is burned.

The energy compartments are:

gsb: green shoot biomass;

gh1, gh2, gh3: the total biomass of nymphs and adults of three major grasshopper species *Aeropedellus clavatus, Melanoplus femurrubrum,* and *Encoptolophus sordidus costalis,* distinguished ecologically by nymph emergence on 15 May, 10 June, and 30 June, respectively [RV1, H];

spm, spf: the biomass of respectively adult male and adult female clay-colored sparrows;

se0: the biomass of developing embryos within the female sparrows;

se1: the biomass of sparrow eggs in incubation;

spnest: the biomass of sparrow nestlings;

spfled: the biomass of sparrow fledglings.

We shall define our units of energy and energy flow as oven dry and in joules per square meter-hour. Energy flows will be computed hourly beginning at hour 0 on 15 May (approximately the times of green shoot biomass emergence and emergence of early grasshopper nymphs).

Even though we plan to model only a portion of one summer, our model must be a sequence of models fitted together over time, adjusting for the appearance and departure of energy compartments. The times of juncture shall be as given in the following chart.

day	15 May	20 May	25 May	1 June	10 June	18 June	30 June	15 July
hour	0	120	240	408	624	816	1104	1464
gsb	--------	--------	--------	--------	--------	--------	--------	--------
gh1	--------	--------	--------	--------	--------	--------	--------	--------
gh2					--------	--------	--------	--------
gh3							--------	--------
spm		--------	--------	--------	--------	--------	--------	--------
spf			--------	--------	--------	--------	--------	--------
se0			--------					
se1				--------				
spnest					--------			
spfled							--------	--------

The intervals shown are averages of times in the literature.

9.2 Accumulation Modeling

Clay-colored sparrows are known to be proficient predators of grasshoppers when breeding. In early summer grasshoppers can be "their chief item of food" [R]. Moreover, grasshoppers constitute "because of their apparent numbers and size, the most important group of invertebrates residing above the soil surface" [RV1]. Hence we shall make the simplifying assumption that sparrows consume nothing but grasshoppers or other arthropods with equivalent energetic value. Finally, we shall assume that all the individuals in each grasshopper species can be taken at will by sparrows. Hence we lump all the stadia of one grasshopper species into one compartment.

Through most of the previous chapters and thoughout classical predator-prey theory, repeated use is made of a "collisions" type energy transfer rate from prey to predator. This approach probably is appropriate to ecosystems in which predator density responds promptly to prey availability, as is the case when yeast "prey upon" sugar molecules in a fermentation vat (Chapter Four).

Collisions type energy transfer seems generally inappropriate, however, for ecosystems with strong seasonal factors and bursts of spring production. Winter in prairie Canada is generally regarded as a season of steady ecosystem energy losses. Many compartments are set to zero by winter. By contrast, it is not generally possible for animals to exploit fully spring food bursts; an abundance of food might lead to saturated demand at many consumer trophic levels. In fact, making as much reproductive use as possible of food bursts is probably a central consideration in animal strategy and design in the temperate zones.

Sparrows nesting at two pairs per hectare cannot possibly have a significant impact on a typical grasshopper density of, say, 10 to 100 grasshoppers per square meter. The collisions type hypothesis for the rate of energy flow from the grasshopper compartment to the sparrow compartment would imply that doubling the grasshopper density would immediately double the energy flow rate through predation; in fact, no such doubling could occur. So an alternative approach is needed.

The alternative to be used here is *accumulation modeling*; it uses known net long-term energy gain values to estimate hourly rates of energy gain. In other words, energy flow due to predation is assumed to be a product

(a function of time of day and year)·(predator density)

as opposed to

(a function of time of day and year)·(predator density)·(prey density)

The time function will be chosen to match net energy accumulation in special circumstances. Such estimates of predation energy flow rates, although not based on what might be called first principles, do lead in turn to a model which can be used to predict ecosystem response to various perturbations and initial conditions.

9.3 Estimating Energy Flows

There are three main vegetation types present in the plant-grasshopper-sparrow system, namely, upland shortgrass (*Stipa* spp. = needle grasses), wheat grasses (*Agropyron* spp.), and pasture sagewort (*Artemisia frigida*); all are represented in the green shoot biomass compartment. Because grasshoppers are indiscriminating as well as wasteful, consequences of plant preference by different grasshopper species (or, indeed, of different stadia within a species) will be assumed to be negligible. We simply assume the green shoot biomass compartment loses energy due to grasshopper predation at a rate determined by time of day and the current sum of grasshopper nymph and adult densities gh1+gh2+gh3. Also, we assume each grasshopper compartment gains energy due to predation on gsb at a rate determined by time of day and the current level of that compartment.

We shall estimate grasshopper wastefulness by assuming that four times as much gsb is dropped as is injested and that about 20% of injested gsb is actually metabolised [BR, L, Waldbauer quoted in RM].

We shall write numbers in "E" notation, so, for example, 1.2E3 represents 1200. The energy worth of all autotroph tissue is assumed to be 2E4 j/gm. The energy worth of all animal tissue is assumed to be 3E4 j/gm (an average of values in [BR]).

Photosynthesis, competition among autotrophs, and energy flow from autotrophs to grasshoppers (grasshopper feeding activity) shall all be assumed to proceed at a current rate equal to a peak rate multiplied by a circadian factor. As a circadian factor we shall use the function

$$f_1(h) = ((1-\cos(\pi h/12)/2)^5$$

where h is the hour since the beginning of May 15 (Fig. 9.1). This factor is simply a mathematical way of incorporating well known circadian rhythms into a model. Note that $f_1(h)$ is positive, that $f_1(h)$ reaches a peak value of 1 at noon each day, and that $f_1(h)$ is very nearly zero overnight (between 1800 hours and 600 hours). Using calculus one can show that the average value of $f_1(h)$ over a 24 hour period is approximately 0.25 . Thus converting an average flow of the above type to a peak flow at noon is achieved to a good approximation by multiplication by four.

177

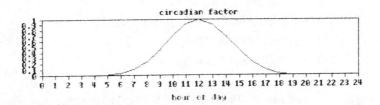

Figure 9.1. Values of a circadian factor, a function used to turn on and off various circadian energy flows in the model.

9.4 Equations and Trajectories

First let us describe a hypothetical model of the upland prairie as it would function for one summer in the absence of animals.

Considering measurements of green shoot biomass in [C], we can deduce the magnitude of peak photosynthesis energy uptake and the effect of competition for nutrients (such as water) in a logistic growth equation. We find

$$gsb(t+\Delta t) = gsb(t) + f_1(h)[.012{\cdot}gsb(t)-.012{\cdot}gsb^2/6E6]\Delta t \qquad (1a)$$

Thus equation (1) reflects the fact that gsb in the absence of animals rises from about 1.2E6 (joules per square meter) to about 6E6 in the time 0 hours to 1464 hours.

The next layer of complexity is the introduction of gh1, gh2, and gh3. We shall asssume that energy flow from gsb to gh1, gh2, gh3 is proportional to current levels of gh1, gh2, and gh3. We also assume that grasshoppers in the nymph or adult stadia injest .5 times their current mass each day and that the net accumulation rate of energy in a grasshopper compartment is .2 times the current injestion rate [approximations of measurements in BR, RM, RV2]. Using also data on grasshopper densities from [RV1], we are led to

$$gsb(t+\Delta t) = gsb(t) + f_1(h)[.012{\cdot}gsb -.012{\cdot}gsb^2/6E6 -(2.5/6){\cdot}\{gh1+gh2+gh3\}]\Delta t$$
$$(1b)$$

The term {gh1+gh2+gh3} must be reduced to {gh1} until h = 625 and to {gh1+gh2} until h = 1105, corresponding to the emergence of gh1 and gh2.

At the same time the equations for gh1, gh2, gh3 are

$$gh1(t+\Delta t) = gh1(t) + f_1(h)[(.1/6)gh1-(1E-6)gh1^2+\{-.02gh1\}] \Delta t \qquad (2a)$$
$$gh2(t+\Delta t) = gh2(t) + f_1(h)[(.1/6)gh2-(1E-6)gh2^2] \Delta t \qquad (3a)$$
$$gh3(t+\Delta t) = gh3(t) + f_1(h)[(.1/6)gh3-(2E-6)gh3^2] \Delta t \qquad (4a)$$

Equation (2a) is initiated at h = 0 with the term {-.02gh1} appearing only after h = 816. Equation (3a) is initiated at h = 625 and equation (4a) is initiated at h = 1105. All grasshopper energy levels are set initially at 1E3 (approximately consistent with data in [RV1]); this could be easily modified to account for egg density variation from year to year in our model.

Additional mathematical complexity is required in order to model energy flows to sparrow compartments. Sparrow feeding activity is confined to and is roughly constant in the daylight hours of each day. Thus we shall employ a switch function $f_2(h)$ which is 1 between 6AM and 6PM and 0 otherwise. Heat loss from adult sparrows has been estimated to be .4 watt per bird at 20° C and .65 watt at 10° C [KDG]. We assume that temperature on average is 20° C at noon each day, 10° C at midnight, and otherwise a function like the circadian function with such maximum and minimum values, that is, $.9-.25f_1$w/bird = $.0108(1-.278f_1)$spm. To balance this loss, net feed gain must be $.0202f_2$spm. Thus to fit mass and energy data from the literature [F, K, R, S] we use

$$spm(t+\Delta t) = spm(t) + [(.00202f_2(h) +.0108(1-.278f_1(h)))\cdot spm] \Delta t \qquad (5)$$

Equation (5) is initiated at h = 120 with an initial value spm = 60 (two males per hectare).

Female sparrows are assumed to arrive at h = 240 with initial spf = 60 (two females per hectare). Females must consume considerable energy to produce embryo biomass, se0. In fact, se0 must increase (approximately linearly) from 0 to 1.2 gm/egg = 3.6E4 joule/egg, four eggs per female, in 168 hours. We shall assume the net conversion efficiency from spf biomass to se0 is 20%. It follows that the equation for spf during the interval h = 240 to h = 408 is

$$\text{spf}(t+\Delta t) = \text{spf}(t) + [(.0202f_2(h) +.0108(1-.278f_1(h)) +.0286\{2f_2(h)-1\})\cdot\text{spf}] \Delta t \tag{6}$$

Equation (6) is initiated at h = 240 and the term +.0286{2f$_2$(h)-1} is deleted after h = 408.

In the equation for se0 we assume that the rate of biomass accumulation is constant, namely .17 j/ha-hr or .00286 j/hr-spf. Thus

$$\text{se0}(t+\Delta t) = \text{se0}(t) + [.00286\cdot\text{spf}] \Delta t \tag{7}$$

Equation (7) is initiated at h = 240 with initial se0 = 0. After h = 408 se0 reaches 28.8 and we terminate (7) and reset se0 = 0 (the eggs are laid).

The equation for se1 is simple. At h = 408 the contents of compartment se0 are moved to compartment se1. The level of se1 remains constant until h = 625 when the contents of se1 are moved to compartment spnest. Hence

$$\text{se1}(t+\Delta t) = \text{se1}(t) \tag{8}$$

for h from 408 through 624. After h = 624 we terminate (8) and set se1 = 0 (the eggs hatch).

Over the period h = 624 to h = 816, we assume eight nestlings grow from 1.2gm to 10 gm each [F, K, S]. Thus spnest grows from 28.8 j/m^2 to 240 j/m^2. In effect we will assume energy flows directly from grasshopper compartments to spnest. That is, we will assume the energy cost of hunting by adults for juveniles is negligible; we do so because the energy cost of elevating a 10 gm bird 10 metres (which must be equivalent to several captures) is only about one joule, a negligible amount compared to heat loss. Thus the equation for net growth of spnest must be

$$\text{spnest}(t+\Delta t) = \text{spnest}(t) + [2.2f_2(h)]\Delta t \tag{9}$$

After h = 816 we terminate (9) and set spnest = 0 (the nestlings leave the nest and become fledglings). At h = 816 the energy content of compartment spnest is transferred to compartment spfled. Each of the eight fledglings per hectare has about the same mass as an adult. Therefore, it is assumed that a fledgling and an adult have about the same energy requirement. It follows that after h = 816 the equation for spfled is

$$\text{spfled}(t+\Delta t) = \text{spfled}(t) + [(.0202f_2(h) +.0108(1-.278f_1(h)))\cdot\text{spfled}] \qquad (10)$$

Now we must return to the equations for gh1, gh2, gh3 and include the various effects of sparrow predation. We find

$$\text{gh1}(t+\Delta t) = \text{gh1}(t) + f_1(h)[(.1/6)\text{gh1}-(2E-6)\text{gh1}^2+\{-.02\text{gh1}\} -5\cdot p1(\text{current sparrow sum})]\ \Delta t \qquad (2b)$$

$$\text{gh2}(t+\Delta t) = \text{gh2}(t) + f_1(h)[(.1/6)\text{gh2}-(2E-6)\text{gh2}^2 -5\cdot p2(\text{current sparrow sum})\ ]\ \Delta t \qquad (3b)$$

$$\text{gh3}(t+\Delta t) = \text{gh3}(t) + f_1(h)[(.1/6)\text{gh3}-(2E-6)\text{gh3}^2 -5\cdot p3(\text{current sparrow sum})\ ]\ \Delta t \qquad (4b)$$

where p1, p2, p3 denote the current proportion of gh1, gh2, gh3 nymphs or adults among all nymphs or adults and flown and flowf denote the net current energy flows to nestlings or fledglings. Note the factor 5 corresponds to a net uptake efficiency by nestlings or fledglings of 20%.

These equations can be implemented in Lotus® 1-2-3®, a popular computer spreadsheet normally used for financial as opposed to ecological bookkeeping. The trajectories of a simulation are shown in Fig. 9.2.

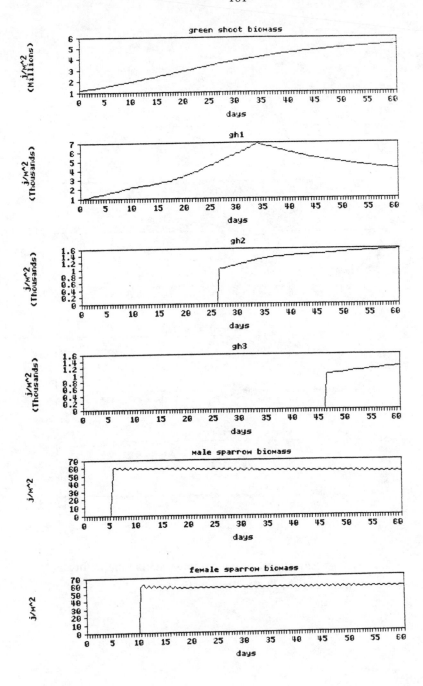

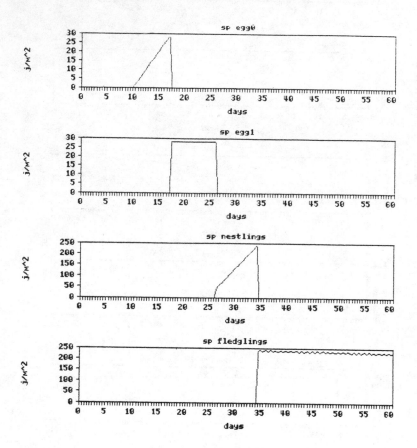

Figure 9.2. Energy trajectories of ten compartments functioning as described by equations 1b, 2b, 3b, 4b, 5,...,10.

9.5 Stability

In contrast to mainstream mathematical ecology theory (meaning idealized models without seasonal variations of photosynthesis and other basic seasonal factors), stability is not at all a feature of our autotrophs, grasshoppers, and sparrows model. This is quite a change from two-dimensional predator-prey models! But neither is stability a feature of the actual shortgrass prairie during the season 15 May--15 July. If the summer season photosynthesis rates could somehow be protracted in a laboratory setting, some compartments would probably expand far above natural levels until overwhelmed by downstream compartments. Then one compartment after another would collapse.

In the real shortgrass prairie, compartment density bounds are maintained by the seasonal disappearance of energy flow pathways. The role of winter is generally to set some compartments back to zero and many others to much reduced levels. In other words, every summer is a time of unstable and erratic energy flow bursts and every winter stabilizes the system by returning compartment levels to spring values with little or no dependence on summer accumulations. It seems possible that many temperate ecosystems are so stabilized by winter.

It turns out to be possible to simulate ecosystem collapse in another way. Modifying the model by increasing initial grasshopper densities or increasing the estimate of grasshopper wastefulness can bring about the obliteration of green shoot biomass by the end of summer. In nature, exactly the same can be the consequence of severe grasshopper infestation.

Acknowledgments. I thank J. F. Doane, D. G. Forsyth, C. F. Hinks, F. A. Leighton, and M. K. Mukerji for their essential and helpful comments. The model represented in this Chapter has been adapted from a preliminary model developed in the course of a research project coordinated by D. G. Irvine (Toxicology Research Centre, Saskatoon, Canada) on the Effects of Grasshopper Insecticides Upon Valuable Wildlife. That project was funded by the Wildlife Toxicology Fund (a co-operative grant program of World Wildlife Fund, Environment Canada, and Noranda Inc.) with matching support from Saskatchewan Wildlife Foundation, Saskatchewan Natural History Society, Saskatchewan Government, Agriculture Canada, and the Toxicology Research Centre. Additional funding for the study was received from the pest fund of Environment Canada.

REFERENCES

[BR] C. Bailey and P. Riegert, Energy dynamics of *Encoptolophus sordidus costalis* (Scudder) (Orthoptera: Acrididae) in a grassland ecosystem, Can. J. Zoo. 51 (1973) 91-100.

[BT] C. H. Buckner and W. J. Turnock, Avian predators on the larch sawfly, *Pristiphora erichsonii,* Ecology 46 (1965) 223-236.

[C] R. Coupland, Tech. Rep. 27, Matador Project, U. of Sask. and NRC, 1973.

[F] Glen A. Fox, A contribution to the life history of the clay-colored sparrow, Auk 78 (1961) 220-224.

[KDG] S. C. Kendeigh, V. R. Dol'nik, and V. M. Gavrilov, Avian energetics in Granivorous Birds in Ecosystems, ed. J. Pinowski and S. C. Kendeigh, Cambridge U. Press, Cambridge, 1977.

[K] Richard W. Knapton, Breeding ecology of the clay-colored sparrow, Living Bird 17 (1979) 137-158.

[L] G. Langford, Some factors relating to the feeding habits of grasshoppers with special reference to *Melanoplus bivittatus.* Bull. Colorado Agric. Exp. Sta. 354 (1930) 1-54.

[RM] R. Randell and R. More, Tech. Rep. 59, Matador Project, U. of Sask. and NRC, 1974.

[RV1] P. Riegert and J. Varley, Tech. Rep. 16, Matador Project, U. of Sask. and NRC, 1973.

[RV2] P. Riegert and J. Varley, Tech. Rep. 17, Matador Project, U. of Sask. and NRC, 1973.

[R] Oscar M. Root, contribution on *Spizella Pallida* in Life Histories of North American Cardinals, Grosbeaks, Buntings, Towhees, Finches, Sparrows, and

Allies, U. S. Nat. Museum Bull. 237, part 2 (1968) 1186-1208

[S] W. Ray Salt, A nesting study of *Spizella Pallida,* Auk 83 (1966) 274-281.

[SL] Yu. M. Svirezhev and D. O. Logofet, *Stability of Biological Communities,* Mir, Moscow, 1983

APPENDIX

Below is a Lotus ® 1-2-3 ® program for computing the trajectories in Fig. 2. Before running the program cells A40..A51 must be copied into cells A40..DR51 and cells B10..B19 must be copied into cells A10.A19. Then enter 1 (= one hour = Δt) in cell B4. Cells A1..A3 constitute a macro labelled \M. The trajectory values are then recorded in cells A42..DR51. The x-axis for graphic output is 0, 5,...,60 (days) in cells A41, K41,...,DR41. The keystrokes "alt" + "m" (or, on some installations, "control"+"shift"+"m") start the model.

```
A1: '{calc}
A2: '/xi(B6-B5)<0~/xgA1~
A3: '/xq
A4: 'delta t (hrs)
B4: 0
E4: (F4) ((1-@COS(@PI*B6/12))/2)^5
F4: (S2) '= diurnal bell
A5: 'max t
B5: 1464
E5: @IF(+B6+6-24*@INT((B6+6)/24)>11,1,0)
F5: (S2) '= diurnal switch
A6: 'current t
B6: +B4+B6
E6: 0.012
F6: (S2) '= r for gsb
A10: (S2) 1200000
B10: (F2) +B30
C10: (S2) 1200000
D10: (F2) '= gsb
E10: 0.5
F10: (S2) '= injest/mass-day
A11: (S2) 1000
B11: (F2) +B31
C11: (S2) 1000
D11: (S2) '= gh1
E11: 0.2
F11: (S2) '= uptake efficiency
A12: (S2) 0
B12: (F2) +B32
C12: (S2) 0
D12: (S2) '= gh2
E12: 5
F12: (S2) '= cons/injest
A13: (S2) 0
B13: (F2) +B33
C13: (S2) 0
D13: (S2) '= gh3
E13: 0.01
F13: (S2) '= eng loss factor
A14: (F0) 0
B14: (F2) +B34
C14: (F0) 0
D14: (F0) '= sfem
A15: (F0) 0
B15: (F2) +B35
C15: (F0) 0
D15: (F0) '= sm
A16: (F0) 0
B16: (F2) +B36
C16: (F0) 0
D16: (F0) '= se0
A17: (F0) 0
B17: (F2) +B37
C17: (F0) 0
D17: (F0) '= se1
A18: (F0) 0
```

```
B18: (F2) +B38
C18: (F0) 0
D18: (F0) '= sn
A19: (F0) 0
B19: (F2) +B39
C19: (F0) 0
D19: (F0) '= sfled
A20: +A10+B4*E4*(E6*A10-E6*A10^2/6000000-4*(E10/24)*E12*(A11+A12+A13))
B20: (F2) '= gsb
C20: (F0) +B6
A21: +A11+B4*E4*(4*(E10/24)*E11*A11-0.000001*A11^2+@IF(B6>816,-E13*A11,0)-
B21: (S2) '= gh1
C21: (S2) +A11/(A11+A12+A13)
D21: (S2) '= proportion gh1
A22: @IF(B6=624,1000,0)+A12+B4*E4*4*((E10/24)*E11*A12-0.000001*A12^2)-5*C2
B22: (S2) '= gh2
C22: (S2) +A12/(A12+A13+A11)
D22: (S2) '= proportion gh2
A23: @IF(B6=1104,1000,0)+A13+B4*E4*4*((E10/24)*E11*A13-0.000001*A13^2)-5*C
B23: (S2) '= gh3
C23: (S2) +A13/(A13+A11+A12)
D23: (S2) '= proportion gh3
A24: @IF(B6=240,60,0)+A14+B4*(2*0.0101*E5*A14-0.0108*(1-0.278*E4)*A14+@IF(
B24: (F0) '= sfem
C24: 2*0.01005*E5*A14+@IF(B6<409,1,0)*0.0143*(2*E5)*A14
D24: (S2) '= gh into sfem
A25: @IF(B6=120,60,0)+A15+B4*(2*0.01005*E5*A15-0.0108*(1-0.278*E4)*A15)
B25: (F0) '= sm
C25: 2*0.01005*E5*A15
D25: (F0) '= gh into sm
A26: (+A16+B4*0.17)*@IF(B6>240,1,0)*@IF(B6<409,1,0)
B26: (F0) '= se0
A27: (+A17+@IF(B6=408,A16,0))*@IF(B6<625,1,0)
B27: (F0) '= se1
A28: @IF(B6=624,A17,0)+@IF(B6>624,1,0)*@IF(B6<817,1,0)*(A18+B4*2.2*E5)
B28: (F0) '= sn
C28: @IF(B6>624,1,0)*@IF(B6<817,1,0)*2.2*E5
D28: (S2) '= gh into sn
A29: @IF(B6=816,A18,0)+A19+B4*(2*0.01005*E5*A19-0.0108*(1-0.278*E4)*A19)
B29: (F0) '= sfled
C29: 2*0.01005*E5*A19
D29: (F0) '= gh into sfled
A30: (F0) +A20
A31: (F0) +A21
A32: (F0) +A22
A33: (F0) +A23
A34: (F0) +A24
A35: (F0) +A25
A36: (F0) +A26
A37: (F0) +A27
A38: (F0) +A28
A39: (F0) +A29
A40: (F0) 0
B40: (F0) 12
C40: (F0) 24
D40: (F0) 36
```

```
E40: (F0) 48
F40: (F0) 60
G40: (F0) 72
H40: (F0) 84
A41: (F0) 0
A42: @IF($B$6=A40,$A$10,A42)
A43: @IF($B$6=A40,$A$11,A43)
A44: @IF($B$6=A40,$A$12,A44)
A45: @IF($B$6=A40,$A$13,A45)
A46: @IF($B$6=A40,$A$14,A46)
A47: @IF($B$6=A40,$A$15,A47)
A48: @IF($B$6=A40,$A$16,A48)
A49: @IF($B$6=A40,$A$17,A49)
A50: @IF($B$6=A40,$A$18,A50)
A51: @IF($B$6=A40,$A$19,A51)
```

INDEX

193